INTRODUCTION TO

SCIENTIFIC COMPUTATION

A First Course for Physics, Mathematics and Engineering Majors

INTRODUCTION TO SCIENTIFIC COMPUTATION

A First Course for Physics, Mathematics and Engineering Majors

J David Brown

North Carolina State University, USA

NEW JERSEY • LONDON • SINGAPORE • BEIJING • SHANGHAI • HONG KONG • TAIPEI • CHENNAI • TOKYO

Published by

World Scientific Publishing Co. Pte. Ltd.
5 Toh Tuck Link, Singapore 596224
USA office: 27 Warren Street, Suite 401-402, Hackensack, NJ 07601
UK office: 57 Shelton Street, Covent Garden, London WC2H 9HE

Library of Congress Cataloging-in-Publication Data
Names: Brown, J. David, 1957- author
Title: Introduction to scientific computation : a first course for physics, mathematics and engineering majors/ J David Brown, North Carolina State University, USA.
Description: Hackensack, New Jersey : World Scientific, [2026] | Includes index. | Contents: Getting started. Integrated development environment -- Jupyter notebook -- Python distribution.
Identifiers: LCCN 2025020906 | ISBN 9789819815487 hardcover | ISBN 9789819816644 paperback | ISBN 9789819815494 ebook for institutions | ISBN 9789819815500 ebook for individuals
Subjects: LCSH: Numerical analysis--Data processing--Textbooks | Numerical analysis--Computer programs--Textbooks | Python (Computer program language)--Problems, exercises, etc.--Textbooks | Physics--Data processing--Textbooks | Engineering--Data processing--Textbooks
Classification: LCC QA297 .B76 2026
LC record available at https://lccn.loc.gov/2025020906

British Library Cataloguing-in-Publication Data
A catalogue record for this book is available from the British Library.

For any available supplementary material, please visit
https://www.worldscientific.com/worldscibooks/10.1142/14375#t=suppl

Desk Editors: Soundararajan Raghuraman/Joseph Ang

Typeset by Stallion Press
Email: enquiries@stallionpress.com

Preface

The seeds for this book were planted in 2014 when I began teaching a course in scientific computation to physics, math and engineering students at NC State University. Those lecture notes underwent countless revisions over the years, culminating in the chapters of this book.

My goal is to present the foundations of scientific computing at a level appropriate for university students in science, engineering and mathematics. The text is intended to be complete; no prior programming experience is needed. Only a solid understanding of calculus is required.

The first nine chapters cover some basic elements of the Python programming language, the elements that are needed for scientific computing. Chapter 10 provides an introduction to symbolic computation. The remaining chapters present numerical techniques for solving the types of problems that commonly appear in science, engineering and math: root finding, curve fitting, interpolation, numerical integration, linear algebra, numerical differentiation, ordinary and partial differential equations, and Fourier analysis.

Scientific computing is fun and interesting. I have tried to infuse that spirit throughout this book. Much of the material is presented in the context of specific problems ranging from Euler–Bernoulli beam theory to the expansion of the Universe. Examples and exercises are drawn from diverse topics such as cellular automata, Kirchoff's laws, the Mandelbrot set, normal mode analysis, Newton's law of cooling, etc. The driven, damped pendulum is treated in detail, providing the reader with valuable experience and insight into numerical methods for differential equations.

This book is intended to be interactive. Exercises are scattered throughout each chapter, and the reader is encouraged to tackle the exercises as they work through the material. Most of the exercises are designed to clarify and reinforce the material as the material is presented, not afterwards.

Errors are present in every numerical calculation. Errors are emphasized throughout the text. Most programming platforms, including Python, provide "canned" routines for solving integrals, differential equations, etc. These built-in routines tend to hide the numerical errors from the user. For this reason, I have chosen to de–emphasize canned routines. In most chapters, numerical algorithms and errors are discussed before any built-in routines are presented. Examples such as the driven, damped pendulum show that canned routines are not always reliable.

Most students in my computational science class have not yet had a course in differential equations. Many have not seen Fourier analysis, and some have had little more than a brief introduction to linear algebra. This is not an issue. In fact, students often benefit from having their first exposure to advanced topics come in a numerical setting. Mathematics is less abstract when treated numerically. Students can gain a hands–on familiarity and understanding of mathematics when they start with a numerical approach. With this foundation in place, they are well prepared to address the more formal aspects of advanced mathematics in their future classes.

I want to thank the members of the committee who helped develop the original content for our course on scientific computing: John Blondin, Carla Fröhlich, James Kneller, Davide Lazzati and Brandon Lunk. I am also grateful for the help and insights provided by those who have taught the course: David Fallest, Christopher Kolb and most recently Karen Daniels and Lex Kemper. Finally, I want to thank Thomas Baumgarte for helpful suggestions and Andrew Brown and Michael Brown for sharing their knowledge and expertise with the Python programming language.

J.D. Brown

Contents

Chapter 1

Getting Started

Scientific computing covers a wide range of topics. This book provides an introduction to some of the basics: integration, differentiation, root finding, linear systems, eigenvalue problems, least squares fitting, ordinary and partial differential equations, interpolation, Fourier analysis, visualization, and symbolic computation. Scientific computing gives you the power to analyze and explore mathematical models of complex physical systems.

The algorithms discussed in this book can be implemented in almost any computer language. We will use Python because Python is powerful, easy to learn, well-supported, and free.

Your first task is to get Python running. You can download the basic Python language from *Python.org*, but this is not the best option. Most scientists and engineers access Python in one or both of the following ways: through an *integrated development environment* or in a *Jupyter notebook*.

1.1 Integrated development environment

An integrated development environment (IDE) is a computer application that provides tools for software development. Some IDEs are designed specifically for Python, others are multi–language. Some IDEs are free, others are not.

A Python program (or code) consists of one or more text files containing Python language statements. An IDE will include a text editor for creating these text files. The text editor will have helpful features such as syntax highlighting, auto completion and real-time error analysis. The program is executed from within the IDE. Some IDEs include an integrated debugger and version control.

The software landscape is changing rapidly. At the time of writing, a good choice for an IDE is *Spyder*. It is free, open source, and is designed for use by scientists and data analysts.

1.2 Jupyter notebook

A Jupyter notebook consists of a mixture of text and Python code, separated into *cells*. This allows Python programs to be organized into groups of commands and interspersed with textual comments and explanations. The text is formatted using Markdown, which allows for figures, tables, LaTeX formulas and other features.

Jupyter notebooks run in your web browser. The actual Python code can be executed locally, on your machine, or in the cloud. At the time of writing, a good choice for a Jupyter notebook application is *JupyterLab*. It is free and open source.

1.3 Python distribution

A *Python distribution* brings together a group of applications that allow the user to develop and execute Python code in a variety of ways. At the time of writing, the distribution *Anaconda* is popular among the science community. It is free for individual users. Anaconda includes *Spyder* and *JupyterLab*, as well as libraries that extend Python's capabilities for scientific computing.

1.4 Python console

IDEs as well as some Jupyter notebook applications include a *Python console*. A Python console (sometimes called an *IPython console*) allows the user to input and execute Python commands one at a time. This gives us a third way to interact with Python, in addition to IDEs and Jupyter notebooks.

1.5 Python as a calculator

Python can be used as a calculator to evaluate simple arithmetic expressions.

Exercise 1.5a

- Enter `2 + 3` in a Python console and type enter (or return).
- Open a text editor in an IDE and type `print(2 + 3)`. Run the program by selecting the run command from a drop-down menu, or by clicking on an appropriate icon. You might need to save the file first.
- Create a Jupyter notebook and type `2 + 3` in a code cell. Then type shift-enter (or shift-return).

When you type `2 + 3` and enter (or return) in a Python console, Python responds with `5`. When you type `2 + 3` in a code cell of a Jupyter notebook, then type shift-enter (or shift-return), Python responds with `5`. The IDE text editor works a bit differently. Python doesn't assume that you want a response unless you tell it so, explicitly, using the `print()` function. When you run the program `print(2 + 3)` Python will "print" the result `5` to the Python console.

Exercise 1.5b

Type `2 + 3`, without the `print()` function, into the IDE editor and run the program. What happened? Python computed `2 + 3` but didn't bother to tell you the answer.

Exercise 1.5c

Compute `2 - 3`, `2*3`, and `2/3`. Use all three methods: a Python console, an IDE text editor, and a Jupyter notebook. Remember, when you run these commands from the text editor you will need to include `print()` to see the results.

You can run a group of commands in a single cell of a Jupyter notebook. In this case, Python will only print the result of the last command.

Exercise 1.5d

In a single code cell of a Jupyter notebook, type

```
2 + 3
2 - 3
```

using the enter (or return) key to separate the two commands. Run the code with shift-enter (or shift-return). The output should be `-1`.

In a Jupyter notebook, just like a program created with an IDE editor, you can tell Python explicitly which results to display using the `print()` function. For example,

```
print(2 + 3)
print(2 - 3)
```

will display both answers, `5` and `-1`.

Python automatically understands basic arithmetic operations, but not more sophisticated mathematical functions like sin and log. In the next chapter we will extend Python's knowledge of mathematics by importing the NumPy library.

Chapter 2

Python Basics

2.1 Exponentiation

In Python, `2` raised to the power `3` is denoted `2**3`, *not* `2^3`.

Exercise 2.1

Compute `2**3` and `2^3`. (Include `print()` if needed.)

The caret symbol ^ is a "bitwise XOR operator." You don't need to understand this right now. Just remember that exponentiation uses the `**` notation.

2.2 Order of operations and parentheses

We can string together calculations—for example,

```
3*4+5**2
```

yields the result `37`. Python follows the standard order of operations: powers come first, then multiplication and division, then addition and subtraction. It is often wise to use parentheses to make sure your expressions are interpreted in the way that you intend. For example, the above is equivalent to

```
(3*4)+(5**2)
```

Parentheses can help make your expressions easier to read.

Exercise 2.2

Use Python to compute `(4*2)+(3*5)` and `4*(2+3)*5`. Are the answers correct? What result does Python give for `1+5*3**6/2`? Express this statement more clearly using one or more sets of parentheses, and check your results.

2.3 Integer division

The sum, difference, or product of two integers is always an integer. For example, $7+4$, $7-4$, and $7 \cdot 4$ yield the integers 11, 3, and 28, respectively. The same is not true for division. We expect the result of 7/4 to be the real number 1.75.

The way computers store numbers into memory depends on whether the number is real (such as 7.0 or 4.0) or integer (such as 7 or 4). As a result, in many programming languages, the character `/` between two integers will return an integer. For example, `7/4` might evaluate to `1`, which is the integer part of 1.75. Of course, in any programming language, real number division should work as expected with `7.0/4.0` giving `1.75`.

Older versions of Python (before version 3.0) treated integer division in this way, with `7/4` evaluating to `1`. Most likely you are using a newer version of Python that treats integer division the way you expect, with `7/4` yielding `1.75`.

The concept of "integer division," where the quotient of integers n and m is defined as the integer part of n/m, can be useful. Newer versions of Python define such an operation by `//`. The related operator `%` gives the remainder.

Exercise 2.3

Evaluate the expressions `7/4`, `7//4`, and `7%4`. What are the results? Replace the integers `7` and `4` with real numbers `7.0` and `4.0`. Do the results change?

2.4 NumPy

The core Python language is somewhat limited, but its vocabulary can be extended by the addition of *libraries*. For scientific computing we often use a library called NumPy (short for "numerical Python"). NumPy can be imported with this command:

```
import numpy
```

NumPy defines a number of constants; for example, π and e are given by

```
numpy.pi
numpy.e
```

NumPy also defines common mathematical functions such as the square root, trig functions, exponentials and logarithms.

Exercise 2.4

Import NumPy and evaluate the following statements.

```
numpy.sqrt(2.0)
numpy.sqrt(25.0)
numpy.sin(30)
numpy.sin(numpy.pi)
numpy.cos(45)
numpy.cos(numpy.pi/2)
numpy.exp(1.0)
numpy.exp(3.0)
numpy.log(numpy.e)
numpy.log(numpy.e*numpy.e)
numpy.log10(10)
numpy.log10(100)
```

- Do NumPy trig functions use radians or degrees?
- What is the base of the `numpy.log()` function?
- What is the base of `numpy.log10()`?

An alternative to `numpy.exp(2.0)` is `numpy.e**2.0`. Similarly, we could write `2.0**0.5` in place of `numpy.sqrt(2.0)`. Most experienced programmers avoid the `**` notation in these contexts and use the NumPy functions `exp()` and `sqrt()` instead.

2.5 Strings

A character is anything that you can type on a keyboard. A *string* is a sequence of one or more characters. Strings (or text) are always surrounded by single or double quotation marks. For example, `'cat'` and `"tree"` are strings.

Observe that `"17.3"` is a string, whereas `17.3` is a number. Python interprets `"17.3"` as a sequence of characters, like any other string. Python interprets `17.3` as a real number.

Strings can be used with the `print()` function; for example,

```
print("Did the cat climb the tree?")
```

Strings are often used to explain the output of a code, as in this example:

```
print("ln(10) = ")
print(numpy.log(10))
```

Exercise 2.5

Run the code above. Modify your code to combine the two print statements into one:

```
print("ln(10) = ", numpy.log(10))
```

2.6 Variables and the = operator

In Python, as in most computer languages, the character = is an *assignment operator*. It tells the computer to evaluate whatever is on the right-hand side then assign the result to the *variable* named on the left-hand side. For example,

```
x = 3 + 5
print("x = ", x)
```

The first line instructs the computer to evaluate `3 + 5`, which is `8`, and to assign `8` to the variable `x`. The second line instructs the computer to print the string `"x = "` and the contents of the variable `x`.

Here is another example:

```
fruit = "banana"
print("fruit = ", fruit)
```

The first line assigns the string `"banana"` to the variable `fruit`. The second line prints the string `"fruit = "` and the contents of the variable `fruit`.

In the world of ordinary mathematics we refer to the character = as an "equals sign." Consider the following statements:

```
3 = 1 + 2
x + y = 5 + 2
z*3 = 15
```

Exercise 2.6a

Execute the above in Python. What are the results?

These statements are perfectly fine in ordinary mathematics but are not allowed in Python. This is because `3`, `x + y`, and `z*3` are not variables.

Variable names in Python can only contain letters, numbers and the underscore character, and they must begin with a letter or an underscore (not a number.) Thus, `xyz_123` is a valid variable name but `123_xyz` is not. Also note that Python is case sensitive. For example, `fruit` and `Fruit` and `fruiT` are all different variable names.

Certain words have special meanings in Python, and cannot be used as variables. These *keywords* include `for`, `if`, `else`, `while`, `break`, `continue`, `and`, `or`, `not`, `return`, and others.[1]

Remember, the assignment operator = should have a variable on the left and an expression to evaluate on the right. The "expression to evaluate" can be a mathematical expression like `3 + 5`. It can also be a single number like `8`, or a single string like `"banana"`.

[1]Python will let you know if you try to use a keyword as a variable name. In scientific computing the only keyword that you might be tempted to use as a variable name is `lambda`.

As an assignment operator, the character `=` can be used in ways that would not make sense in ordinary mathematics. For example, the code

```
x = 4
x = x + 7
```

is perfectly fine in Python. The first statement assigns the number `4` to the variable `x`. The second statement evaluates `x + 7`, which is `11`; Python then assigns the result `11` to the variable `x`, replacing the previous value in the process.

Exercise 2.6b

Execute the above in Python. Insert the statement `print(x)` at various places in the code. Does `x` equal `4` or `11`?

Exercise 2.6c

Assume that `x` has been assigned the value `13`. Which of these Python statements is valid? Invalid? Why?

- `x2x = x*2*x`
- `x = x + "banana"`
- `banana = "3 + 5"`
- `y = z = 27`
- `4 = 4`
- `y = 6 = 13`
- `apple&orange = "fruit"`

2.7 Projectile motion

A projectile near Earth follows a trajectory in the x–y plane (the x-axis is horizontal and the y-axis is vertical) given by

$$x(t) = x_0 + v_{0x}t, \tag{2.1a}$$

$$y(t) = y_0 + v_{0y}t - \frac{1}{2}gt^2. \tag{2.1b}$$

Here, x_0, y_0 are the components of the initial position, v_{0x}, v_{0y} are the components of the initial velocity, and g is the acceleration due to gravity. Of course t is time. Let's write a Python program to calculate the position of the projectile at $t = 2.0\,\text{s}$, given the initial conditions $x_0 = 0.0\,\text{m}$, $y_0 = 5.0\,\text{m}$, $v_{0x} = 10.0\,\text{m/s}$ and $v_{0y} = 15.0\,\text{m/s}$:

```
# Position of a projectile near Earth at time t

x0 = 0.0        # x-component of initial position (m)
y0 = 5.0        # y-component of initial position (m)
v0x = 10.0      # x-component of initial velocity (m/s)
v0y = 15.0      # y-component of initial velocity (m/s)
g = 9.8         # acceleration due to gravity (m/s**2)
t = 2.0         # time (s)

# Compute the position at time t
x = x0 + v0x*t
y = y0 + v0y*t - 0.5*g*t**2

# Print the results
print("Position at t = ", t, "seconds:")
print("x = ", x, "meters")
print("y = ", y, "meters")
```

This program can be created in an IDE text editor, or in one or more code cells of a Jupyter notebook.

Observe the following features of the program.

- Comments are used throughout. Any text that follows `#` is a comment. Adding comments to your code will make it easier for other people to understand. Comments will help you as well—as time goes by it can be difficult to remember the details and logic of your own codes.
- Blank lines are inserted between sections of code. We begin with a description of the code, then a section where variables are defined. The position at time t is computed in the next section, and the results are printed in the final section. Separating the code into sections makes it easier to read.
- Numbers are assigned to variables. This is useful because we might want change the numbers in the calculation. For example, we can easily change the time from `2.0` to `3.0` by modifying the single statement `t = 2.0`. There is no need to search through the code for every occurrence of the number `2.0`, and then decide if it should be changed to `3.0`.

- The variable names correspond closely to the variables used in the formulas (2.1). This makes the variables easy to recognize. For example, the x-component of initial velocity is denoted v_{0x} in the formulas and `v0x` in the code. Another good name would be `v_0x`.
- The print statements include text strings that explain the output.

Exercise 2.7

Run the code to compute the position of the projectile at various times. Experiment with different values for the initial position and velocity.

2.8 User input

We can change the value of time in our projectile code by modifying the statement `t = 2.0`. Another option is to let the user choose the value of `t` when the code is executed. This can be done with the `input()` function. Simply replace the statement `t = 2.0` with

```
t = input("Choose time t: ")
t = float(t)
```

When the program reaches the line with `input()`, it pauses and waits for user input. Python interprets the user input as a string, and assigns that string to the variable `t`. In the next line, the function `float()` converts the string `t` into a real number (also called a "float").

You can prompt the user for multiple inputs. In place of the assignment statements for `t`, `v0x`, and `v0y` in the projectile code, we could write

```
t, v0x, v0y = input("Choose t, v0x, v0y: ").split()
```

then use `float()` to convert each variable `t`, `v0x` and `v0y` into a real number. The user's input values must be separated by spaces.

Exercise 2.8

Modify your projectile code to include user input for time `t` and the components of velocity, `v0x` and `v0y`.

Remember, `input()` interprets user input as a string. If the string is to be treated as a number, it must be converted to a real number or an integer. The function `float()` converts a string of digits with or without a decimal point into a real number. The function `int()` converts a string of digits without a decimal point into an integer.

2.9 Output formatting

Let's ask Python to compute 5.0/3.0 and print the result:

```
x = 5.0/3.0
print("five-thirds =", x)
```

We can achieve the same output with this code:

```
x = 5.0/3.0
print(f"five-thirds = {x}")
```

The `f` tells Python that what follows is an *f-string* (or *formatted string–literal*), which can include variables in curly brackets. The curly brackets can also contain expressions such as `x**2` or `3*x`.

The output of each code above is

```
five-thirds = 1.6666666666666667
```

We might not want to see so many digits. The output can be displayed in a more readable form by adding formatting instructions:

```
x = 5.0/3.0
print(f"five-thirds = {x:.3f}")
```

This yields the output

```
five-thirds = 1.667
```

The expression `.3f` after the colon tells Python to display `x` as a real number rounded to 3 decimal places. (The `f` stands for "float," another name for a real number.)

Exercise 2.9

Modify the `print()` statements in your projectile code so that the results are displayed to 5 decimal places.

2.10 Lists

A Python *list* is a collection of numbers or strings surrounded by square brackets:

```
mylist = [1.0, 'elephant', -12.3, 3]
```

Lists can be added, for example,

```
secondlist = [44, -7.6, 'tiger']
biglist = mylist + secondlist
print(biglist)
```

Later, we will learn other ways to manipulate lists.

Exercise 2.10

Create two lists in Python and add them. Can you multiply a list by an integer? Can you multiply two lists?

2.11 Line continuation and indentation

Most Python commands occupy a single line of code. If a command is long, it might be easier to read if it is extended across multiple lines. You can tell Python that a command extends to the next line using the line continuation character \. Note that \ should be the last character on the line that is to be continued. Make sure there are no spaces after \.

Exercise 2.11a

Consider the code

```
x = 2 + 3
print(f"x = {x}")
```

which outputs `x = 5`. Place a line continuation character \ after the `+` sign, and move the number `3` to the next line. Rerun the code.

Python can sometimes recognize that a command is incomplete and must extend to the next line. Typically, an opening left parenthesis `(` without a closing right parenthesis `)` signals an incomplete command that will be continued on the next line. Likewise, an opening square bracket `[` without a closing square bracket `]` tells Python that the command is incomplete. In these cases line continuation characters are not needed.

Exercise 2.11b

Modify the code above by placing parentheses around `2 + 3`. Break the command into two lines, somewhere between the parentheses. Does the code run? Now replace the statement `x = 2 + 3` with `x = [2, 3]`, turning `x` into a list. Extend the list across two lines, splitting it after the comma. Does the code run?

In later chapters, we will see that some Python commands need to be indented. (None of the commands discussed so far should be indented.) For commands that are spread across multiple lines, the first line must be indented properly. Indentation for the subsequent lines doesn't matter—you can choose any pattern of indentation that makes your code easy to read.

Exercise 2.11c

Some programmers like to spread out the elements of a list; for example

```
x = ["cat",
     "dog",
     "fish",
     "bird"]
print(x)
```

Run this code and experiment with different amounts of indentation.

2.12 Order of execution

Consider the sequence of Python commands

```
x = 3.0
y = x + x**2
print(y)
```

If you run this code from an IDE editor, the commands will be executed in order, one after the other: `x = 3.0` is executed first, then `y = x + x**2`, then `print(y)`. Likewise, you could enter this code into a single cell of a Jupyter notebook. When the cell is executed, the commands will be executed in the order in which they appear. Alternatively, you might spread these commands across two or more cells of a Jupyter notebook. In this case the contents of each cell are executed when you type shift-enter (or shift-return). This is not necessarily the order in which the cells appear in the notebook.

For example, you could place

```
y = x + x**2
print(y)
```

in the first cell of a Jupyter notebook, and

```
x = 3.0
```

in the second cell. Although this is confusing and not recommended, the results will be fine if the second cell is executed before the first cell.

Exercise 2.12

Place `y = x + x**2` and `print(y)` in the first cell of a Jupyter notebook, and `x = 3.0` in the second cell. Execute the second cell (shift–enter), then the first. What happens? Now restart Python (often referred to as "restarting the kernel") and execute the two commands in the opposite order (the order in which they appear). What happens?

Chapter 3

Control Structures

3.1 The `for` loop

Control structures allow sections of your program to be repeated or skipped when certain conditions are met. One of the most useful control structures is the `for` loop. Here is the syntax:

> `for` *variable* `in` *list*:
>   *code to execute*

This is best illustrated with a simple example:

```
for x in [1.2, 3.5, "cat"]:
    print(x)
```

Here, the *variable* is `x` and the *list* is `[1.2, 3.5, "cat"]`. The *code to execute* is `print(x)`. The `for` statement tells Python to execute the indented code for each value of the variable `x`. The *code to execute* is repeated three times, first, for `x = 1.2`, then for `x = 3.5`, then for `x = "cat"`.

Note that the `for` statement always ends in a colon and *code to execute* is indented. The indentation can be any number of spaces, or the tab character. If the *code to execute* requires multiple commands, then each line must be indented. Any unindented statements that follow the indented lines of code will not be included in the `for` loop.

Exercise 3.1a

Run the code above. Make sure to indent the print statement. Add other numbers or strings to the list. Add one more line, such as `print("wow!")`, to the end of the code. What happens if this last command is indented? Not indented?

Here is an example that uses a `for` loop to calculate 4!, the factorial of the number 4:

```
fac = 1                  # assign the variable fac to 1

for i in [2, 3, 4]:      # repeat indented code for i = 2, 3, 4
    fac = i*fac

print(f"4! = {fac}")     # print the result
```

The `for` statement causes the indented line to repeat for each of the three values of the variable `i` that appear in the list `[2, 3, 4]`. This code is logically equivalent to

```
fac = 1                  # assign the variable fac to 1

i = 2                    # set i = 2
fac = i*fac              # reassign fac to 2*1 = 2

i = 3                    # set i = 3
fac = i*fac              # reassign fac to 3*2 = 6

i = 4                    # set i = 4
fac = i*fac              # reassign fac to 4*6 = 24

print(f"4! = {fac}") # print the result
```

Exercise 3.1b

The "double factorial" is defined by

$$n!! = \begin{cases} n \cdot (n-2) \cdot \cdots \cdot 3 \cdot 1\ , & n \text{ odd,} \\ n \cdot (n-2) \cdot \cdots \cdot 4 \cdot 2\ , & n \text{ even.} \end{cases}$$

Write a program that uses a `for` loop to compute 13!!.

Note that indentation must be consistent throughout your program. Whether you choose a certain number of spaces, or the tab character, you must stick with that choice. You cannot mix indentation types within a single code. It does not matter if comments are indented.

Exercise 3.1c

Write a code to compute the square roots of 3, 6, 9, 12 and 15. Use a `for` loop and a list. Print the results to four decimal places.

3.2 The `range()` function

In the last section we computed 4! using a `for` loop along with the list `[2, 3, 4]`. How can we modify our code to compute 97!? It would be tedious to create a list of integers from 2 through 97. This problem is solved with the `range()` function:

```
fac = 1                  # assign the variable fac to 1

for i in range(2,98): # repeat indented code for i = 2,...,97
    fac = i*fac

print(f"97! = {fac})  # print the result
```

The command `range(a,b)` creates a Python list of integers beginning with `a`, and ending with `b-1`. That's right—the last number in the list is `b-1`. The number `b` in `range(a,b)` is the first integer that is *excluded* from the list. In the code above, `range(2,98)` produces the list of integers $2, \ldots, 97$.

With the `range()` function, we can easily modify our code to compute the factorial of any positive integer chosen by the user:

```
# input number and convert the string to an integer
n = input("input a positive integer:")
n = int(n)

# set fac to 1 then repeat indented code for i = 2,...,n
fac = 1
for i in range(2,n+1):
    fac = i*fac

print(f"{n}! = {fac})
```

Exercise 3.2a

Write a program that asks the user to input two integers, n and m, with $n > m$. The code should compute the product

$$m \cdot (m+1) \cdots (n-1) \cdot n.$$

(This is the product of all integers beginning with m and ending with n.) Test your code with small input values to make sure it works correctly.

Exercise 3.2b

Modify the projectile code from Sec. 2.7 to compute the projectile's position at times $t = 1, 2, \ldots, 5$.

3.3 More about `range()`

The `range()` function produces a list of integers.[1] We can use this function in various ways. For integers `a`, `b`, `s`, and `n`,

- `range(a,b)` creates a list of integers beginning with `a`, incrementing by 1, and ending *before* `b` (that is, ending with `b-1`).
- `range(a,b,s)` creates a list of integers beginning with `a`, incrementing by `s`, and ending *before* `b`.
- `range(n)` creates a list of n integers beginning with 0, incrementing by 1, and ending *before* `n` (that is, ending with `n-1`).

Exercise 3.3a

Run this code:

```
for i in range(-2,5):
    print(f"i = {i}")
```

[1]Strickly speaking, `range()` creates an "immutable sequence," not a list. (Immutable means the elements cannot be modified.) An immutable sequence behaves like a list when it appears in a `for` loop.

```
for j in range(2,8,3):
    print(f"j = {j}")

for k in range(4):
    print(f"k = {k}")
```

Pay attention to the first and last numbers in each list!

Exercise 3.3b

Write a program that asks the user to input an odd integer n, then computes $n!!$. Test your code with small input values to make sure it works correctly.

The following code uses a `for` loop and `range()` to compute

$$\sum_{n=1}^{10} n,$$

the sum of the first 10 positive integers:

```
sum = 0
for n in range(1,11):
    sum = sum + n
print("sum = ", sum)
```

Note that the variable `sum` is given an initial value of `0` before entering the loop.

Exercise 3.3c

Write a code to compute

$$S = \sum_{k=1}^{N} k = 1 + 2 + 3 + \cdots + N$$

for a given integer N. Experiment with different values of N to find the smallest number N such that $S > 10^6$.

3.4 Conditions

A *condition* is a statement that is either `True` or `False`. Most conditions involve a comparison between two quantities. For example, the condition `5 > 8` is `False`. You can use the following operators for comparison:

== (equivalence, equal to)
!= (not equal to)
< (less than)
> (greater than)
<= (less than or equal to)
>= (greater than or equal to)

Exercise 3.4a

Predict the output from this code:

```
x = 3 + 2
print(x == 5)
y = 4 + 5
print(y == "cat")
```

Now run the code. Were your predictions correct?

Remember: The symbol = is an "assignment operator" and == is a comparison operator.

Exercise 3.4b

Evaluate these conditionals:

- 0 == 5
- 0 <= 5
- 0 > 5
- 6 != 3
- 3 == 3
- 6 != 6
- 2 < 2
- 2 >= 2

3.5 if, elif, else

Conditions can be used with the `if`, `elif`, `else` construct to control the logical flow of a code. The keyword `elif` is short for "else if."

The `if` statement can be used alone:

> `if` *condition*:
>
> *code to execute if the condition evaluates to* `True`

If *condition* evaluates to `True`, then Python executes the indented line (or lines) of code below the `if` statement. If *condition* evaluates to `False`, Python skips the indented line (or lines) of code.

The `if` statement can be used with `else`:

> `if` *condition*:
>
> *code to execute if the condition evaluates to* `True`
>
> `else`:
>
> *code to execute if the condition evaluates to* `False`

The `if` statement can also be used with `elif` and `else`:

> `if` *condition 1*:
>
> *code to execute if condition 1 evaluates to* `True`
>
> `elif` *condition 2*:
>
> *code to execute if condition 1 evaluates to* `False` *and condition 2 evaluates to* `True`
>
> `else`:
>
> *code to execute if both condition 1 and condition 2 evaluate to* `False`

Note that, just as with a `for` statement, the blocks of code to be executed are indented. Also there is a colon : following each of the `if`, `elif` and `else` statements. Observe:

- You can extend the `if` construction to include as many `elif` branches as you like.
- You can have only one `if` branch and one `else` branch.
- The `elif` and `else` branches are optional.

Let's assume the user has been asked to input two numbers, M and N. The code

```
if M > N:
    print("M is greater than N")
```

will check to see if the number M is greater than N.

Exercise 3.5a

Extend this code to determine whether $M > N$, $M < N$ or $M = N$ by using `if` and two `elif` branches. Modify your code to use `if` with one `elif` branch and one `else` branch.

`if` statements can appear anywhere in a code, including the middle of a loop. In the following example, the `for` loop is executed 10 times, once for each value of n from 1 through 10. Inside the loop the `if` and `else` statements check to see if the numbers n are even or odd. Recall that `n%2` gives the remainder of n divided by 2.

```
for n in range(1,11):
    if n % 2 == 0:          # n is even
        print(n,'is even')
    else:                   # n is odd
        print(n,'is odd')
```

Observe that the code to execute under the `for` statement is indented. The codes to execute under the `if` and `else` statements are "doubly indented."

Exercise 3.5b

Write a program that asks the user to input any positive integer n (even or odd) then computes $n!!$.

Exercise 3.5c

Write a code that will examine every integer n from 1 through 100. If the integer is divisible by 7, the code should print n and n^2. If the number is divisible by 11, the code should print n and $\sqrt{n}$.

3.6 `and`, `or`, `not`

Sometimes more than one condition needs to be satisfied. The expression

condition 1 `and` *condition 2*

will evaluate to `True` if both *condition 1* and *condition 2* are `True`. The expression

condition 1 `or` *condition 2*

will evaluate to `True` if either *condition 1* or *condition 2* (or both) are `True`. The command

`not` *condition*

will evaluate to `True` if *condition* evaluates to `False`.

As an example, you might like to know which numbers from 1 through 20 are divisible by 2 and by 3:

```
for n in range(1,21):
    if n%2 == 0 and n%3 == 0:
        print(n, 'is divisible by two and by three')
```

Exercise 3.6

Find the numbers 1 through 20 that are divisible by either 2 or 3 or both.

3.7 `continue` and `break`

The keywords `continue` and `break` are used to alter the flow of the code through a `for` loop. These keywords can also be used with `while` loops; see Sec. 7.6.

Imagine a code that asks the user to choose a positive integer N. The code then computes the squares of each integer from 1 through N. After the integer has been input, the code looks like this:

```
for i in range(1,N+1):
    isqr = i*i
    print(i,'*',i,'=',isqr)
print('the end')
```

For $N = 4$, the output is

```
1 * 1 = 1
2 * 2 = 4
3 * 3 = 9
4 * 4 = 16
the end
```

Now imagine a code that asks the user to input two positive integers, N and M. The code computes the squares of each integer from 1 through N, but skips M. Modify the code above by adding `continue` inside an `if` statement:

```
for i in range(1,N+1):
    if i == M:
        continue
    isqr = i*i
    print(i,'*',i,'=',isqr)
print('the end')
```

For $N = 4$ and $M = 3$, the output is

```
1 * 1 = 1
2 * 2 = 4
4 * 4 = 16
the end
```

The `continue` keyword interrupts the normal cycle of the `for` loop. As shown in Fig. 3.1, `continue` stops the current cycle through the loop and instructs the code to continue with the next cycle.

Finally, imagine a code that asks the user to input positive integers N and M. The code computes the squares of each integer from 1 through N, but skips M and all integers larger than M. If we could be sure that $M \leq N$, we could simply change the range of the `for` loop to `range(1,M+1)`. But this won't work if $M > N$. Instead, we can modify the previous code by replacing `continue` with `break`:

```
for i in range(1,N+1):
    if i == M:
        break
    isqr = i*i
    print(i,'*',i,'=',isqr)
print('the end')
```

```
True  ┌─→ for i in range(1,N+1):
      └──────────┬─────────────┐
                 │if i == M:   │
False ┌──────────┤    continue │
      └─────────→└─────────────┘
                  isqr = i*i
                  print(i,'*',i,'=',isqr)
              print('the end')
```

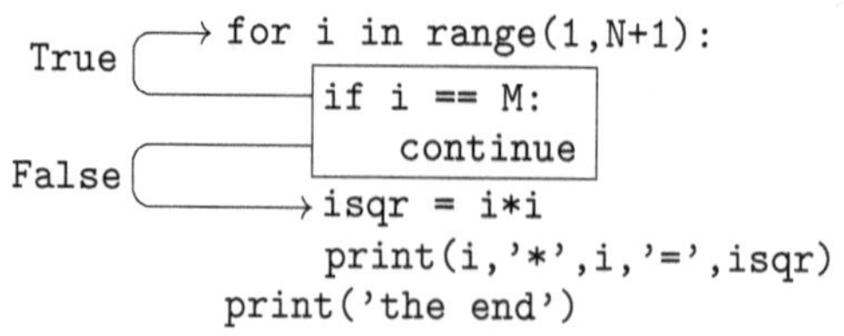

Fig. 3.1. When the condition `i == M` evaluates to `True`, the `continue` keyword tells the code to continue with the next cycle of the `for` loop.

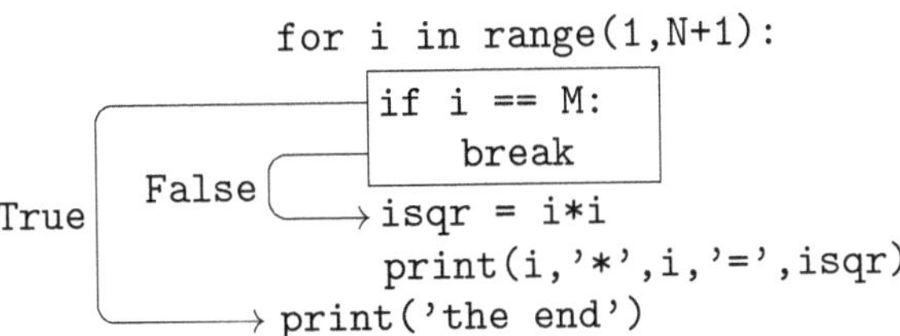

Fig. 3.2. When the condition `i == M` evaluates to `True`, the `break` keyword tells the code to stop cycling through the `for` loop.

For $N = 4$ and $M = 3$, the output is

```
1 * 1 = 1
2 * 2 = 4
the end
```

The `break` keyword also interrupts the normal cycle of the `for` loop. As shown in Fig. 3.2, `break` instructs the code to break away from the `for` loop and move to the next line of code outside the loop.

Exercise 3.7a

Create a code that asks the user to input three positive integers, N, M_1 and M_2. The code should compute (and print out) the squares of all integers 1 through N, but skip M_1 and stop computing if it reaches M_2.

Exercise 3.7b

Write a Python program to compute the sum of all positive integers 1 through 100 that are divisible by 3 or by 5, but not both.

3.8 More exercises

Exercise 3.8a

It is well known that the geometric series

$$\sum_{n=0}^{\infty} 1/2^n$$

is equal to 2. Verify this by computing the partial sum $\sum_{n=0}^{N} 1/2^n$ and showing that the result tends to 2 as N increases.

Exercise 3.8b

Carry out numerical experiments to approximate the value of the Riemann zeta function

$$\zeta(s) = \sum_{n=1}^{\infty} \frac{1}{n^s},$$

for $s = 2$, 3 and 4. Compare with the known results $\zeta(2) = \pi^2/6$ and $\zeta(4) = \pi^4/90$. There is no closed form expression for $\zeta(3)$.

Exercise 3.8c

The Leibniz formula for π,

$$\pi = 4 \sum_{k=1}^{\infty} \frac{(-1)^{k+1}}{2k-1}$$

is obtained by evaluating the Taylor series for $\tan^{-1} x$ at $x = 1$. Approximate the infinite series by computing partial sums. How many terms are required for the answer to match π to 6 decimal places?

Chapter 4

Libraries, Arrays, and Plots

4.1 Importing libraries

Many Python functions are grouped together into libraries that can be imported into a program. Trig functions, exponentials, logarithms, etc., are contained in a library called NumPy. In Chapter 2 we saw how to import and use NumPy:

```
import numpy
# function names must include the prefix numpy
rootseven = numpy.sqrt(7.0)
```

This code computes the square root of 7.0 and assigns the result to the variable `rootseven`.

It can be tiresome to type the prefix `numpy` as part of the name for common mathematical functions. An alternative is to import all NumPy commands with an asterisk:

```
from numpy import *
# function names don't need a prefix
rootseven = sqrt(7.0)
```

This works, but can be dangerous. If we import more than one library in this way, we might have two different functions with the same name. This can lead to conflicts.

The solution is to import the library with a shorter name, like this:

```
import numpy as np
# function names must include the prefix np
rootseven = np.sqrt(7.0)
```

Experienced programmers consider this to be the best way to import a library.

4.2 NumPy arrays

Recall that a Python *list* is a collection of numbers or strings. For example, `[8, -3.2, "tree"]` is a list. A NumPy *array* is similar to a list, but differs in the way it's stored in computer memory. NumPy arrays are used almost exclusively for numerical computation. We will only work with NumPy arrays whose elements are numbers, not strings.

One way to create a NumPy array is with the `linspace()` function. The array `linspace(a,b,N)` consists of `N` evenly spaced numbers beginning with `a` and ending with `b`. Note that `N` must be an integer. The numbers `a` and `b` can be either integer or real. For example, consider the code

```
import numpy as np
myarray = np.linspace(1.0, 3.0, 5)
print(myarray)
```

This array contains five evenly spaced numbers beginning with 1.0 and ending with 3.0.

The output of the code above is

```
[1.  1.5 2.  2.5 3. ]
```

This looks a lot like a list of numbers. For comparison, consider

```
mylist = [1, 1.5, 2, 2.5, 3]
print(mylist)
```

which outputs the list

```
[1, 1.5, 2, 2.5, 3]
```

When printed, both arrays and lists are surrounded by square brackets. However, array elements are separated by spaces whereas list elements are separated by commas.

More importantly, NumPy arrays and Python lists differ in their mathematical behavior:

- Addition of two NumPy arrays occurs on an element-by-element basis. Addition of two lists simply concatenates the lists.
- Multiplication of a NumPy array by a number occurs on an element-by-element basis. Multiplication of a list by an integer n concatenates n copies of the list. Multiplication of a list by a real number is not defined.
- Multiplication of two NumPy arrays occurs on an element-by-element basis. Multiplication of two lists is not defined.

Exercise 4.2a

Create a NumPy array `A = [2. 2.2 2.4 2.6 2.8 3. ]` using the `linspace()` function, and create a Python list `B = [2.0, 2.2, 2.4, 2.6, 2.8, 3.0]`. Find the results of the following calculations:

- `A + A`
- `B + B`
- `3*A`
- `3*B`
- `A*A`
- `B*B`

Note the differences in behavior between NumPy arrays and Python lists.

The advantage of NumPy arrays, as opposed to lists of numbers, is that Python can perform mathematical operations on arrays without cycling over the individual elements one by one. For example, the following program computes the squares of the first five integers:

```
for x in [1,2,3,4,5]:
    xsquared = x**2
    print(x, xsquared)
```

The same result is obtained more efficiently using a NumPy array:

```
import numpy as np
x = np.linspace(1,5,5)
xsquared = x**2
print(x, xsquared)
```

Since `x` is a NumPy array, the command `xsquared = x**2` tells Python to create a new NumPy array, called `xsquared`, whose elements are the squares of the elements of `x`.

Exercise 4.2b

We considered projectile motion in Sec. 2.7. Write a program that computes a projectile's position at times $t = 0.0, 0.25, 0.5, \ldots, 5.0$. Use `linspace()` to create a NumPy array for t. Compute the arrays x and y from Eqs. (2.1) without using a `for` loop. (You can choose any values of initial position and initial velocity that you like.)

4.3 Creating NumPy arrays

There are many ways to create NumPy arrays.

- `linspace(a,b,N)` creates an array with `N` equally spaced elements starting with `a` and ending with `b`.
- `array([a,b,c])` creates a NumPy array with elements `a`, `b`, and `c`.
- `zeros(N)` creates a NumPy array with `N` elements, all zeros.
- `ones(N)` creates a NumPy array with `N` elements, all ones.

Exercise 4.3a

Create NumPy arrays using each of the functions above. Do they give the expected results?

Another way to create a NumPy array is the `arange()` function, which is similar to the `range()` function discussed in Sec. 3.2. In particular,

- `arange(a,b)` creates a NumPy array that begins at `a`, increments by 1, and ends *before* `b`.
- `arange(a,b,s)` creates a NumPy array that begins at `a`, increments by `s`, and ends *before* `b`.
- `arange(b)` creates an array of integers, beginning with 0, incrementing by 1, and ending *before* `b`.

Remember, `arange()` creates a NumPy array, whereas `range()` creates a list. Another difference is this: for `arange()` the parameters `a`, `b`, and `s` can be real numbers; for `range()` the parameters must be integers.

Exercise 4.3b

Create NumPy arrays using `arange()`. Experiment with different real and integer values for `a`, `b`, and `s`. Do they give the expected results?

As a general rule, you should use `linspace(a,b,N)` for creating arrays of real numbers. It can be tricky to use `arange(a,b,s)` if any of the values `a`, `b`, or `s` are not integers.

Exercise 4.3c

Create the following arrays:

```
array1 = np.arange(2,7,1)
array2 = np.arange(13/3,11,2/3)
```

Can you predict the outcome in each case?

The command `arange(a,b,s)` is supposed to *exclude* `b`. Occasionally, when dealing with nonintegers, `arange(a,b,s)` can make a mistake and include `b`. This is due to machine roundoff error.

4.4 Machine roundoff error

Most modern computers are 64-bit systems—they use 64 bits of memory to approximate a real number. One bit is used for the sign of the number and 11 bits are used for the exponent. The remaining

52 bits encode the significand. With this representation, real numbers are approximated to roughly 15, 16, or 17 significant figures, depending on the number.

To be precise, 64-bit computers encode real numbers as

$$\text{real number} = (-1)^s\,(1+A)\,2^{B-1023}, \tag{4.1}$$

where

$$A = \sum_{i=1}^{52} a_i 2^{-i},$$

$$B = \sum_{j=0}^{10} b_j 2^j.$$

Here, s, a_i, and b_j denote bits in memory, either 0 or 1. The single bit s determines the sign of the number. The 52 bits $a_1, \ldots, a_{52}$ comprise the significand, and the 11 bits $b_0, \ldots, b_{10}$ determine the exponent.

As an example, the real number 2/3 is *approximated* by Eq. (4.1) with $s = 0$ and

$$a_i = \begin{cases} 0 & \text{if } i \text{ is odd,} \\ 1 & \text{if } i \text{ is even,} \end{cases}$$

$$b_j = \begin{cases} 0 & \text{if } j = 0 \text{ or } 10, \\ 1 & \text{otherwise.} \end{cases}$$

This number is actually

$$0.66666666666666666296\ldots,$$

slightly less than 2/3. The discrepancy is machine roundoff error.

Exercise 4.4a

Issue the command

```
print(f"x = {x:.20f}")
```

for various real numbers x, such as 2/3. Here, the print function uses f-string formatting to display the results to 20 decimal places. Can 0.3 be represented exactly? 0.1? 0.75?

Due to the finite representation of real numbers, even simple operations are subject to errors.

Exercise 4.4b

Does the conditional

```
1.3 - 1.0 == 0.3
```

evaluate to `True` or `False`?

Machine roundoff errors are present in all numerical calculations. Some consequences of machine error are discussed in later chapters.

Exercise 4.4c

Create a code that assigns variables $x = 0.1$, $y = 0.2$, $z = 0.3$, then computes

```
sum1 = x + (y + z)
sum2 = (x + y) + z
```

Check the equivalence of `sum1` and `sum2` using the comparison operator `==`.

4.5 Plotting graphs

To plot graphs in Python, we need to import the `matplotlib.pyplot` library:

```
import matplotlib.pyplot as plt
```

`plt` is a common abbreviation for `matplotlib.pyplot`.

The code below plots a graph of the function $y = x^3$ from -1 to 1:

```
import numpy as np
import matplotlib.pyplot as plt

# Create a numpy array of x values
x = np.linspace(-1, 1, 100)
# Create a numpy array of y values
y = x**3
```

```
# Graph y versus x
plt.close()                  # close plot windows
plt.plot(x,y,"k.")           # plot y versus x using black dots
plt.title("Function x**3")   # create a plot title
plt.xlabel("x")              # create a label for the x-axis
plt.ylabel("y")              # create a label for the y-axis
plt.show()                   # display the results
```

The result is shown in Fig. 4.1.

The function `close()` closes any plot windows that might have been opened previously. The function `show()` displays the figure as directed in the preceding pyplot commands. `close()` and `show()` can be omitted in Jupyter notebooks, but might be needed on other Python platforms.

Observe that `plot(t,y,"k.")` displays each data point as a black dot. If you replace `"k."` with `"k"`, the graph will be a smooth curve through the data. Replace `k` with `r` for red, `b` for blue, `g` for green, `y` for yellow, `c` for cyan, or `m` for magenta.

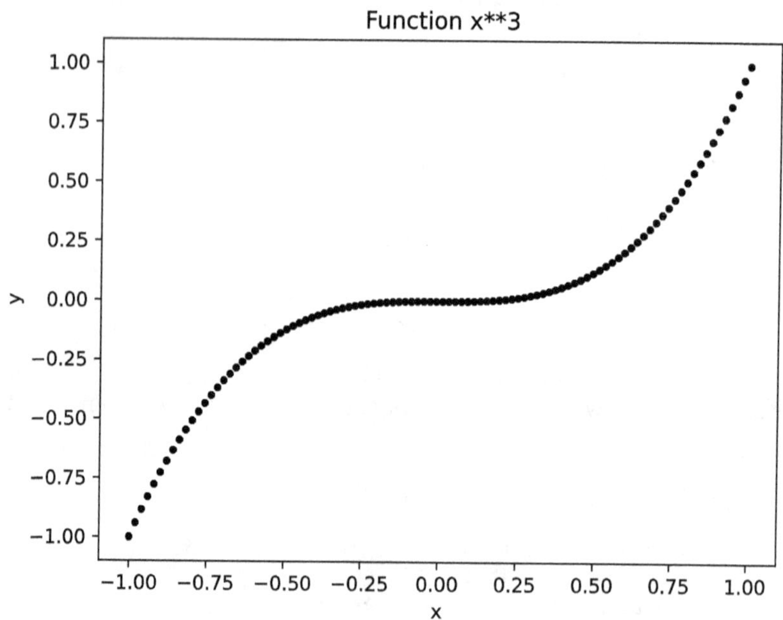

Fig. 4.1. Graph of $y = x^3$ created in Python with Matplotlib.

Exercise 4.5a

Create a new version of your projectile code. Use the `linspace` command to create a time array with 100 or more values. Use a single command (not a `for` loop) to compute the array of x values. Similarly for the y array. Have your code plot the shape of the trajectory, y versus x.

Exercise 4.5b

Plot the functions $f_1(x) = x^3$ and $f_2(x) = x^4$ from -1 to 1. Use `linspace` to create an array of x values, then create arrays `y1 = x**3` and `y2 = x**4`. Plot the two functions on the same graph by using two `plt.plot` commands. You might want to choose different colors for the two curves.

Chapter 5

Indexing Lists and Arrays

5.1 NumPy and Matplotlib

The NumPy and Matplotlib libraries are frequently used in scientific computing. In the example programs below (and in future chapters) these import commands are not shown. We will assume that, if needed, NumPy has been imported as `np` and `matplotlib.pyplot` has been imported as `plt`.

5.2 Lists and arrays

In the last chapter we learned how Python lists and NumPy arrays behave differently.

Exercise 5.2

Create the following lists and arrays:

- `listone = [4,2,5,3]`
- `listtwo = [3,7,9,1]`
- `arrayone = np.array([4,2,5,3])`
- `arraytwo = np.array([3,7,9,1])`

Compute `listone + listtwo` and `arrayone + arraytwo`. What is the difference? Compute `2*listone` and `2*arrayone`. What is the difference? What is the result of `listone* listtwo`? `arrayone*arraytwo`? `listone + arrayone`?

5.3 Indexing

Given a list or array `A` we can reference an individual element by writing `A[i]`, where the index `i` is an integer. The first element in the list or array has index `0`, the second element has index `1`, and so on. As an example, consider the list `A = [2.7, "cat", -8]`. Obviously, `A` has three elements, and we say that the "length" of `A` is three. The individual elements are `A[0] = 2.7`, `A[1] = "cat"` and `A[2] = -8`.

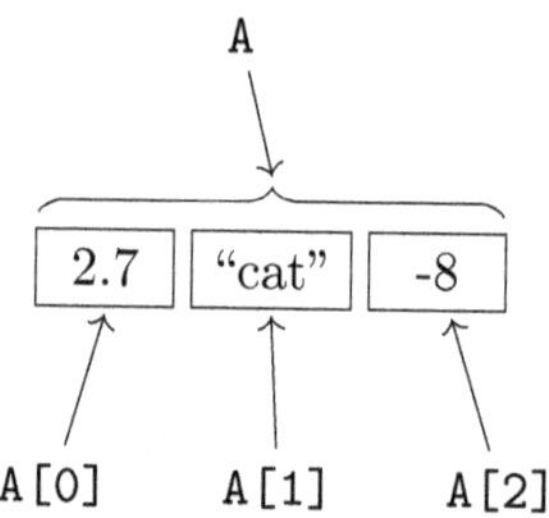

Exercise 5.3a

Consider these lists and arrays:

```
A = [2, 4, 6, 8, 10, 12]
B = np.linspace(0.1, 5.1, 6)
```

- How many elements are in `A`? In `B`?
- What is `A[0]`? `B[0]`?
- What is `A[2]`? `B[2]`?
- What is `A[5]`? `B[5]`?
- What is `A[6]`? `B[6]`?

These examples illustrate a very important aspect of Python's behavior: *Indices for lists and arrays begin with* 0. If there are N elements in a list or array called `A`, the first element is `A[0]` and the last element is `A[N-1]`.

Exercise 5.3b

Run this code:

```
C = np.linspace(2,4,5)
print(C)
C[3] = 17.0
print(C)
```

What's happening?

In the first line, `linspace()` creates the array `[2.0 2.5 3.0 3.5 4.0]` and assigns it to the variable `C`. In the statement `C[3] = 17.0`, the *fourth* array element is changed from its previous value, `3.5`, to the new value `17.0`. The array becomes `[2.0 2.5 3.0 17.0 4.0]`.

5.4 Accessing elements

You can use variables as indices to access the elements of a list or array. Consider the following code:

```
A = np.arange(6,15,1)                # create an array A
print("array length = ", len(A))  # print the length of A

i = 2                                # choose an index value
print("element ", i, "is ", A[i]) # print element i
```

This code uses the function `len()`, which gives the length of a list or array (the number of elements in the list or array).

Exercise 5.4a

Run the code above. Do you understand the output? What happens when you try to access an array element that is greater than or equal to the length of the array?

Another example of accessing array elements:

```
A = np.array([6.1, -15.3, -3.7, 7.4, 20.4, 11.8, -6.5])
N = len(A)
for i in range(N):
    if A[i] < 0:
        A[i] = -A[i]
print(A)
```

This code takes the array `A` and reverses the sign of every element that is negative.

- The first line creates `A`.
- The second line defines `N` as the length of `A`.
- In the third line, the command `range(N)` creates a list of integers beginning with `0` and ending with `N-1`.
- The `for` loop assigns `i` to each element of `range(N)`, one at a time, and executes the indented lines below.
- If the array element `A[i]` is negative, the `if` statement evaluates to `True`. The indented code below the `if` statement reverses the sign of `A[i]`.
- The print function prints the elements of the array with negative elements switched to positive.

Note that we are using `N` rather than 7 inside the `range` command. This helps avoid errors, in particular if we decide to change the number of elements in `A`.

> **Exercise 5.4b**
>
> Write a code that will create a NumPy array consisting of numbers $1, 2, \ldots, 100$. Have your code search through the array using a `for` loop to cycle through the index values. Print the value of every array element that is divisible by 3, but not divisible by 2.

5.5 Using arrays

Let's create an array `A` of real numbers from 0 through 10, spaced by 0.5, then construct a new array containing the square roots of those numbers.

```
A = np.linspace(0,10,21)          # Create a real number array
roots = np.zeros(len(A))          # Create array for square roots

for i in range(len(A)):           # Loop over array indices
    roots[i] = np.sqrt(A[i])      # Compute the square root

print("roots =", roots)           # Print the answer
```

Make sure you understand each line of this code.

- Why is the last argument of `linspace()` equal to 21?
- What values does `i` take in the `for` loop?
- Why did we include the command `roots = np.zeros(len(A))` in the second line?

Take a close look at this last question. The command `roots = np.zeros(len(A))` creates the array `roots` and fills it with 0's. We could create the array using a different NumPy function; for example, we could use `ones()` instead of `zeros()`. The initial values of the array don't matter since those values are overwritten in the `for` loop. Note, however, that the `roots` array should have the same length as `A`.

The real question is this: Why do we need to create the `roots` array at all?

Exercise 5.5a

Run the code above. Eliminate (or comment out) the line `roots = np.zeros(len(A))`, and run the code again. (If you are using a Jupyter notebook, you need to restart the kernel before running the code. This causes Python to forget that `A` is a NumPy array.) What sort of error message do you get?

Without the command `roots = np.zeros(len(A))`, the code doesn't run. Here's why: Without that line, the first appearance of the variable `roots` is in the statement `roots[i] = np.sqrt(A[i])`. Python doesn't know how to interpret `roots[i]` because Python doesn't know that `roots` is an array. It might seem obvious that `roots` is either a list or an array, since it carries an index. But Python isn't able to allocate memory for `roots` since it can't guess what type of object it is, or how many elements it should contain.

So, we need to tell Python that `roots` is an array of length `len(A)` before we can assign values to the elements `roots[i]`. We do this with the statement `roots = np.zeros(len(A))`.

Another way to produce an array of square roots is to use the approach discussed in Sec. 4.2:

```
A = np.linspace(0,10,21)       # Create a real number array
roots = np.sqrt(A)             # Compute square roots
print("roots =",roots)         # Print the answer
```

This code is more efficient than the previous code, and more simple. In particular, there is no need to create the `roots` array using `zeros()` or some other NumPy function. This is because Python already knows that `A` is a NumPy array. When Python sees the statement `roots = np.sqrt(A)`, it can infer that `roots` is an array of length `len(A)`.

Exercise 5.5b

Make an array `A = np.linspace(-10,10,25)`. Use a `for` loop and array indexing to create an array `B` whose elements are the same as `A`, but in reverse order.

Exercise 5.5c

A ball of clay is dropped from an initial height y_0 above the floor. While the ball is in the air, its height as a function of time t is

$$y(t) = y_0 - \frac{1}{2}gt^2,$$

where $g = 9.8\,\mathrm{m/s^2}$ is the acceleration due to gravity. When the ball hits the floor it sticks; its height is then $y(t) = 0.0$. Write a code to plot a graph of y versus t for $0.0 \le t \le 2.0\,\mathrm{s}$, using $y_0 = 2.0\,\mathrm{m}$. One way to do this: create an array for `t` using `linspace()` and an array for `y` using the formula above. Then loop through the elements of `y` and replace all negative values with 0.0.

5.6 Counters

How many numbers from N_1 through N_2 are divisible by 3, but not divisible by 2? For example, with $N_1 = 7$ and $N_2 = 28$ the numbers that are divisible by 3 but not by 2 are these: 9, 15, 21, and 27. The number of such numbers is 4.

The code below solves this problem using a *counter* variable called `ctr`:

```
N1 = 7                  # choose value for N1
N2 = 28                 # choose value for N2
ctr = 0                 # initialize counter to 0
for n in range(N1, N2+1):
    if n%3 == 0 and n%2 != 0:
        ctr = ctr + 1
print(ctr)
```

After assigning values to N_1 and N_2, the variable `ctr` is initialized to 0. In the `for` loop `n` is assigned values from N_1 through N_2, one after the other. The `if` statement determines if `n` is divisible by 3, but not divisible by 2, and if so, increments `ctr` by 1. The variable `ctr` counts the number of times the condition `n%3 == 0 and n%2 != 0` evaluates to `True`.

Exercise 5.6

How many numbers from 1 through 10000 are divisible by 5, but not by 7?

5.7 Appending to an array

Rather than simply counting the number of numbers that are divisible by 3 but not 2, let's collect these numbers into an array `A`. Here is a somewhat unsatisfactory solution to the problem:

```
N1 = 7
N2 = 28
ctr = 0
A = np.zeros(4)       # create array A

# loop over n from N1 through N2
for n in range(N1, N2+1):
    # test to see if n is divisible by 3 but not by 2
    if n%3 == 0 and n%2 != 0:
        A[ctr] = n      # set array element to n
        ctr = ctr + 1   # increment counter
print(ctr, A)
```

This code unsatisfactory because the command `A = np.zeros(4)`, which is used to create `A`, requires prior knowledge that the array `A` should have four elements. What if we change N_1 and N_2? How can we tell how large the array A should be?

There is a better way to structure this code. Begin with an empty array, then use the NumPy append() function to add array elements as they are found.

```
N1 = 7
N2 = 28
A = []                    # create an empty array A

# loop over n from N1 through N2
for n in range(N1, N2+1):
    # test to see if n is divisible by 3 but not by 2
    if n%3 == 0 and n%2 != 0:
        A = np.append(A,n)      # append n to A
print(A)
```

The NumPy command append(A,n) creates a new array with n appended to the end of A. Python then replaces the old A array with the new array. Note that a counter is not needed for this code.

Exercise 5.7a

Write a program that creates an array of numbers from N_1 through N_2 that are divisible either by 3 or by 5, but not both. Use the NumPy append() function. Add a counter to determine the number of such numbers.

Exercise 5.7b

Write a code that produces an array containing the cumulative sum of integers up to a chosen integer N. For example, if $N = 5$, the result should be [1 3 6 10 15]. (These are 1, $1+2$, $1+2+3$, $1+2+3+4$, $1+2+3+4+5$.)

5.8 Appending to a list

Appending an element to a list is not the same as appending an element to a NumPy array. To add an element *newelement* to the end of a list mylist, simply issue the command

```
mylist.append(newelement)
```

Exercise 5.8a

Create a list containing your favorite animals. Start with an empty list and append the elements, one-by-one.

5.9 More exercises

Exercise 5.9a

The Fibonacci sequence is defined by the recurrence relation

$$F_0 = 0,$$
$$F_1 = 1,$$
$$F_n = F_{n-1} + F_{n-2},$$

where n is an integer greater than 1. The first few terms in the sequence are $0, 1, 1, 2, 3, 5, 8, \ldots$. Write a code to compute the Fibonacci sequence up to $n = 100$ and plot $\log(F_n)$ versus n. Find the slope of the curve near $n = 100$. According to Binet's formula, the slope should approach $\log(\varphi)$ for large n, where $\varphi = (1 + \sqrt{5})/2$ is the golden ratio.

Exercise 5.9b

Sterling's approximation to $n!$ is

$$S(n) = \sqrt{2\pi n}\left(\frac{n}{e}\right)^n.$$

Create a graph showing the percent relative error $|1 - S(n)/n!| \cdot 100$ in Sterling's approximation for positive integers n.

Chapter 6

Data Types and Variable Assignment

6.1 Python data types

Python defines various data types, including `int` (integers), `float` (real numbers, also known as *floats*), `complex` (complex numbers), `list` (lists), and `str` (strings). The data type of any object `O` can be found using the command `type(O)`.

The distinction between integers (`int`) and real numbers (`float`) is mostly transparent in Python, but can be important in other programming languages.

Complex numbers (`complex`) are new to us. Python uses `j` for $\sqrt{-1}$. For example, a complex number with real part 2.0 and imaginary part 3.0 is denoted `2.0 + 3.0j`. Note that there is no asterisk (no multiplication sign) between `3.0` and `j`.

When we combine an integer with a real number through addition, subtraction, multiplication or division, the result is a real number. When we combine an integer or a real number with a complex number, the result is a complex number. In Sec. 2.3, we observed that for Python versions 3.0 and beyond, the division of two integers gives a real number.

Exercise 6.1a

Choose an integer `i`, a real number `r`, and a complex number `c`. Compute the sums `i+r`, `i+c` and `r+c` and check their types using the `type()` function. Do the same for multiplication and division. What type is `i/i`?

A list (`list`) is a collection of numbers or strings, such as `L = [2.2, "tree", -4]`. Elements of a list are accessed with the square bracket notation. Element `L[0]` is `2.2`, element `L[1]` is `"tree"`, etc.

A string (`str`) is a sequence of characters surrounded by single or double quotation marks. Thus, `"cat"` is a string. The characters of a string can be accessed with the square bracket notation, just like the elements of a list. For the string `S = "cat"`, element `S[0]` is `c`, element `S[1]` is `a`, etc. Like lists, strings can be added together or multiplied by an integer. For example, `"cat" + "dog"` is `"catdog"` and `2*"cat"` is `"catcat"`.

Exercise 6.1b

Consider the list `L = [2.2, "tree", -4]`. Verify that `L[1]` is `tree` and `L[2]` is `-4`. What is `L[1][2]`?

Exercise 6.1c

Let `C` and `D` denote two strings of equal length. Write a code that will interleave these strings together. For example, if `C = "cat"` and `D = "dog"`, the resulting string should be `"cdaotg"`. Your code should work with other pairs of strings, such as `C = "orange"` and `D = "purple"`.

6.2 NumPy data types

The NumPy library defines more data types, including `matrix` (matrices) and `ndarray` (n-dimensional arrays). We will discuss matrices in the chapter on linear algebra. Objects of the type `ndarray` include the one-dimensional arrays discussed previously. Those arrays are created using NumPy functions such as `linspace()`, `array()` and `zeros()`.

By default, the elements of an `ndarray` are real numbers (floats). You can override the default by specifying the data type when the array is created. For example, the command `np.zeros(10,dtype=int)` will create a NumPy array of 10 integers, initially all 0's.

Exercise 6.2a

Experiment with various data types for NumPy arrays. What is the difference between `np.zeros(10)` and `np.zeros(10,dtype=int)`? What is the difference between `np.linspace(1,4,9)` and `np.linspace(1,4,9,dtype=int)`? What happens if you create an array of integers, then try to change one of its elements to a float?

Here is a two-dimensional NumPy array:

```
A = np.array([[1.0,2.0,3.0],[4.0,5.0,6.0]])
```

Pay close attention to the syntax. The round parentheses are present because `array()` is a function. Inside the round parentheses we have nested sets of square brackets. The inner brackets define the one-dimensional arrays `[1.0,2.0,3.0]` and `[4.0,5.0,6.0]`. The outer brackets define a one-dimensional array whose first element is `[1.0,2.0,3.0]` and second element is `[4.0,5.0,6.0]`. In other words, we can view `A` as an array of arrays.

The elements of an n-dimensional array can be accessed with indices. For example, the second element of `A` (the element with index `1`) is

```
A[1] = [4.0,5.0,6.0]
```

which is itself an array. The third element of the array `A[1]` (the element with index `2`) is

```
A[1][2] = 6.0
```

Another notation for `A[1][2]` is

```
A[1,2] = 6.0
```

When printed, Python displays the array `A` as

```
[[1. 2. 3.]
 [4. 5. 6.]]
```

Thus, we can view `A` as a 2×3 matrix whose rows are `A[0]` and `A[1]`.

Exercise 6.2b

Create two 3×3 NumPy arrays `A` and `B`. Compute:

- `3*A`
- `A + B`
- `A*B`

Note that two-dimensional NumPy arrays do *not* multiply like matrices.

An n-dimensional NumPy array can be created using `zeros()` or `ones()`. For example, `zeros((3,4))` will produce a two-dimensional array of size 3×4, filled with zero's. Note the nested round parentheses.

Exercise 6.2c

Write a program that will ask the user to input a positive integer n, then create an $n \times n$ multiplication table. For $n = 3$, the output should appear as

```
[[1. 2. 3.]
 [2. 4. 6.]
 [3. 6. 9.]]
```

6.3 Lists of lists

A two-dimensional NumPy array can be viewed as an array of arrays. Likewise, we can have a list of lists. In other words, each element of a list can be a list. For example:

```
L = [[-3.3,"cat"], [2.5, "tree"], ["dog",1.7]]
```

is a list whose three elements are `L[0] = [-3.3,"cat"]`, `L[1] = [2.5, "tree"]`, and `L[2] = ["dog",1.7]`. The two elements of `L[1]` are `L[1][0] = 2.5` and `L[1][1] = "tree"`. Unlike a two-dimensional array, we cannot access elements of a two-dimensional list by enclosing the indices in a single pair of square brackets. That is, we can't write `L[1,1]` in place of `L[1][1]`.

Exercise 6.3

The elements of a list can include a mixture of numbers, strings and lists of varying lengths. Modify the code above by replacing `L[2]` with a list of length 3. Also replace `L[1][0]` with a list. How would you access an element of that list?

6.4 Type conversion

We can convert integers and real numbers to strings using the `str()` function:

```
n = 17                      # choose integer
x = 23.1                    # choose real number
print(type(n), type(x))
nstring = str(n)            # convert n to a string
xstring = str(x)            # convert x to a string
print(type(nstring), type(xstring))
```

If a string consists of digits without a decimal point, we can convert it to an integer using the `int()` function. If a string consists of digits with or without a decimal point, we can convert it to a real number using the `float()` function:

```
s1 = "4321"                 # string without a decimal point
s2 = "34.87"                # string with a decimal point
print(type(s1), type(s2))
s1int = int(s1)             # convert s1 to integer
s2float = float(s2)         # convert s2 to real number
print(type(s1int), type(s2float))
```

We previously saw (Sec. 2.8) that the `input` command interprets user input as a string. If the input is to be used as a number, it must be converted using `float()` or `int()`.

Exercise 6.4a

Run the codes above, and observe the output.

How many 0's are in the number 2^{53}? To answer this question, we need to convert the number 2^{53} to a string:

```
i = 2**53                        # Create the integer 2**53
i = str(i)                       # Convert the integer to a string
count = 0                        # Initialize counter
for n in range(len(i)):          # Loop over elements of the string
    if i[n] == "0":              # Check if string element equals 0
        count = count + 1        # Increment the counter by 1
print(count)
```

Exercise 6.4b

Find all powers of 2 through 2^{100} that do *not* contain the digit 3. (Powers of 2 are numbers of the form 2^n with n a positive integer; that is, 2, 4, 8, 16, etc.) How many such numbers are there? What is the largest?

6.5 Variable assignment for numbers

Consider the following scenario. A variable `x` has been assigned to some value. We now set `y = x`. What happens to `y` if we modify `x`? What happens to `x` if we modify `y`?

Exercise 6.5a

Run the program

```
x = 5.1      # assign value to x
y = x        # set y equal to x
x = 3.7      # modify x
print(x, y)
```

What are the final values of `x` and `y`? Are the results what you expected?

Here's how Python interprets this program. We start with a blank section of memory, depicted as follows:

In the first line, `x = 5.1`, Python instructs the computer to write the number `5.1` into memory. The computer then creates the variable `x` that points to that memory location.

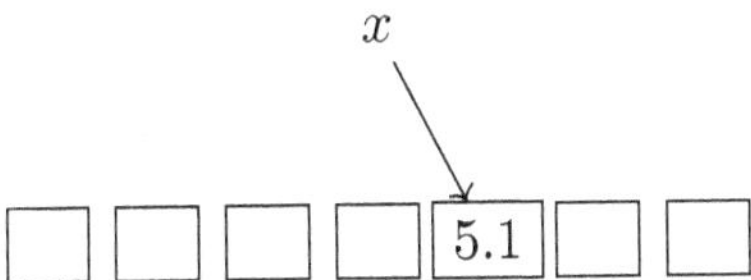

The second line is `y = x`. Python evaluates the right-hand side, which is `5.1`, then creates a new variable `y` that points to the location in memory containing `5.1`.

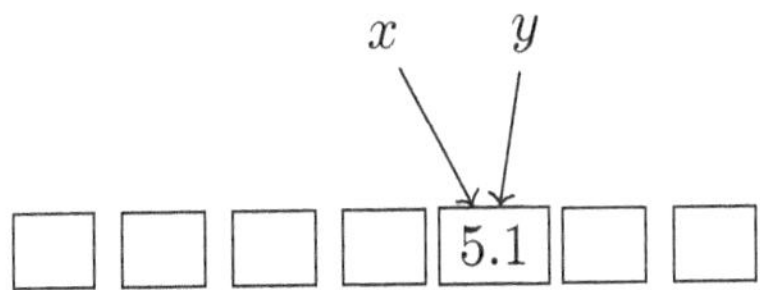

The third line is `x = 3.7`. Python instructs the computer to write the number `3.7` into memory and then redirect the variable `x` to that memory location.

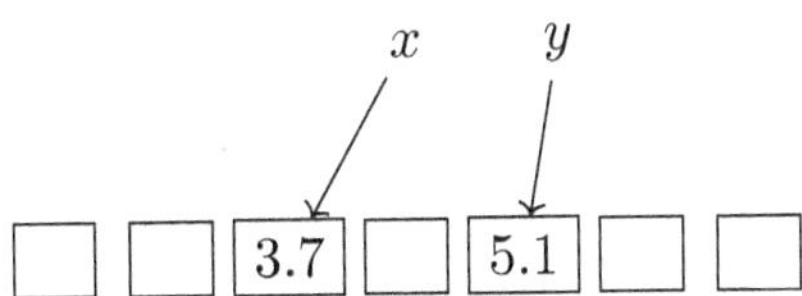

As you can see, the final value of `x` is `3.7` and the final value of `y` is `5.1`. The reassignment of `x` from `5.1` to `3.7` does *not* change the value of `y`. The relation `y = x` no longer holds.

Exercise 6.5b

Predict the output of this program:

```
x = 5.1
y = x
y = 7.4        # modify y
print(x,y)
```

Were you correct?

6.6 Variable assignment for lists and arrays

How does Python interpret `y = x` when the variable `x` is a list or array? Let's analyze the program

```
x = [2.7, "cat", -8]
y = x
x = [-3.3, 17, "cat"]
```

The first line instructs Python to write the list elements, `2.1`, `"cat"` and `-8` into memory, then create a variable `x` that points to that list.

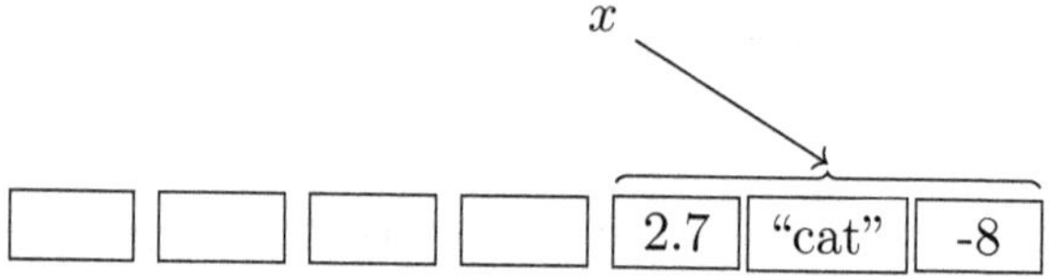

The second line is `y = x`. Python evaluates the right-hand side, which is the list `[2.7, "cat", -8]`, then creates a new variable `y` that points to the memory location containing the list.

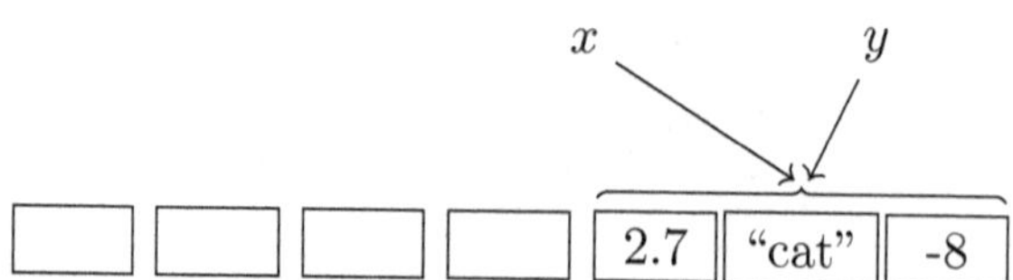

The third line is `x = [-3.3, 17, "cat"]`. The computer writes the list elements `-3.3`, `17` and `"cat"` into memory and then redirects the variable `x` to that memory location.

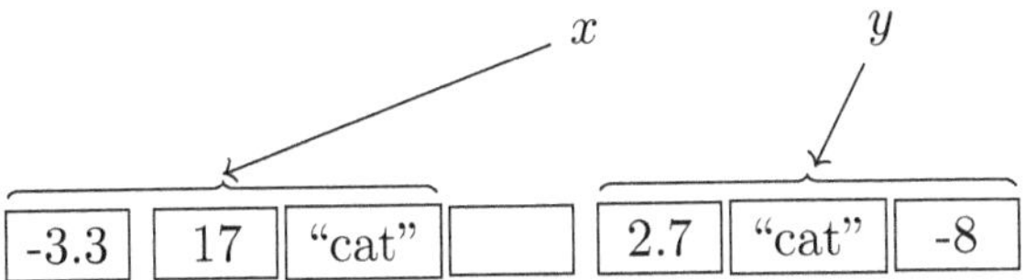

The final value of x is `[-3.3, 17, "cat"]` and the final value of y is `[2.7, "cat", -8]`. The reassignment of x does not change the value of y, so the relation `y = x` no longer holds.

This isn't the end of the story. Consider the program

```
x = [2.7, "cat", -8]
y = x
x[1] = "dog"
```

The first line creates the list `[2.7, "cat", -8]` in memory and points x to that list. The second line points y to the same list.

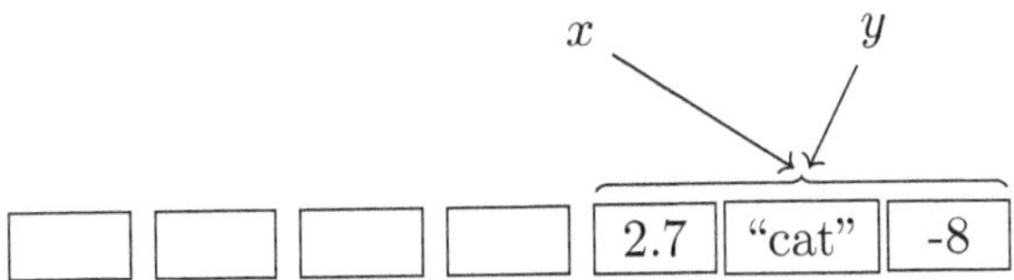

The third line, `x[1] = "dog"`, tells the computer to modify the list by replacing `"cat"` with `"dog"`, The final values for *both* x and y is the list `[2.7, "dog", -8]`.

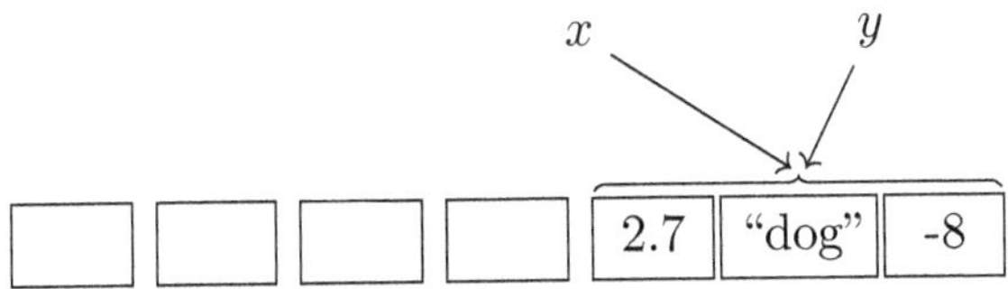

NumPy arrays behave the same way as lists. The statement `x = np.linspace(0.0,4.0,5)` places the array `[0. 1. 2. 3. 4.]` into memory and points x to that array. We can assign a new variable name to that same array using `y = x`. Subsequently,

- We can use the statement `x = np.linspace(10.0,20.0,11)` to redirect x to a new array. This does not change y, so x and y are now different arrays.

- We can use the statement `x[3] = 17.0` to modify the array from `[0. 1. 2. 3. 4.]` to `[0. 1. 2. 17. 4.]`. Both x and y point to the modified array, so x and y are still equal.

Exercise 6.6

Experiment with the program

```
x = np.linspace(0.0, 4.0, 5)
y = x
# modify or replace the x or y array
print(x,y)
```

What happens if you replace every element of x individually? What happens if you replace one or more elements of y? What happens if you replace y with a new array?

6.7 Copying arrays and lists

If x is a NumPy array and you need to create another copy, you probably don't want to use `y = x`. As we saw in the last section, `y = x` simply creates a new name, y, for the array. Both x and y point to the same array. If you modify an individual element of x, this will also modify y. Likewise, if you modify an individual element of y, this will change x.

Rather than `y = x`, you should use the NumPy `copy()` function:

```
x = np.linspace(5.1,7.1,3)  # create an array
y = np.copy(x)              # make a copy of the array
```

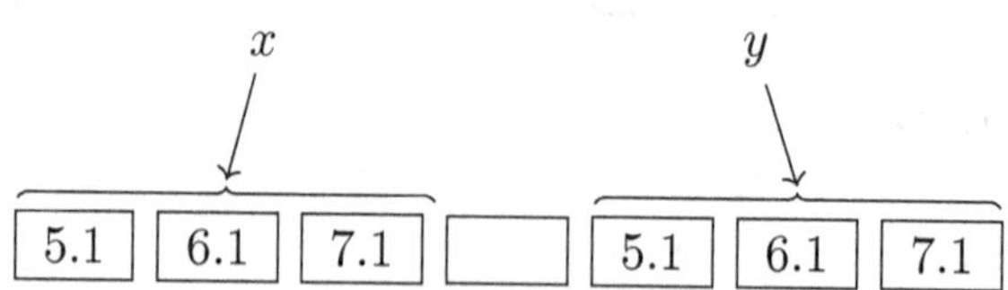

Using `y = np.copy(x)` creates a new array y whose elements are equal to the elements of x. The arrays x and y are independent from one another.

Exercise 6.7a

Create a NumPy array `x` and use `copy()` to copy the array to `y`. What happens to each array if you change the value of `x[1]`? If you change the value of `y[3]`?

The same advice applies to lists: you should avoid using `y = x`. Note that the syntax for copying a list is a bit different from copying a NumPy array:

```
x = [2.7, "cat", -8]  # create a list
y = x.copy()          # copy the list
print(y)
x[1] = "dog"          # change element x[1]
print(x, y)
```

Exercise 6.7b

Run the code above and and check the final results for `x` and `y`. What happens if you replace the line `x[1] = "dog"` with `y[1] = "dog"`?

6.8 Cellular automata

A cellular automaton consists of a regular grid of cells. Each cell can be either *on* or *off*. These two states are represented by +1 (for on) and 0 (for off). The cells evolve, switching their states between on and off, according to some predefined rules.

Consider a simple example with a one-dimensional grid. We can number the cells by an index j. Observe that each cell has two neighbors, one to the left and one to the right. In particular, the jth cell has neighbors $j-1$ and $j+1$. Here is one possible set of rules:

- If the two neighbors of cell j are in different states, switch the state of j.
- If the two neighbors of cell j are in the same state, leave the state of j alone.

The picture below shows the jth cell and its nearest neighbors. Cell $j-1$ is off (0), cells j and $j+1$ are on (+1). Since the neighbors of j

are in different states, cell j will switch from on $(+1)$ to off (0) when the rules are applied.

$$\cdots \boxed{0}\boxed{1}\boxed{1} \cdots$$
$$\cdots \quad j-1 \quad j \quad j+1 \quad \cdots$$

Let's write a code to simulate the evolution of these cells. Begin by creating the grid of, say, 20 cells, all in the off (0) state. Then flip some interior cells to on $(+1)$:

```
C = np.zeros(20,dtype=int)
C[10] = 1
```

Now we implement the rules, repeating five times:

```
for n in range(5):
    D = np.copy(C)
    for j in range(1,19):
        if D[j-1] != D[j+1]:
            C[j] = 1 - C[j]
print(C)
```

Comments:

- The array `C` is created using `dtype=int`, and `D` is a copy of `C`. As a result the elements of `D` are integers. We chose to use integers because we want to avoid machine roundoff errors in the evaluation of the condition `D[j-1] != D[j+1]`.[1]
- The outer `for` loop causes the rules to be repeated five times, once for each value of `n` from 0 through 4. The `print(C)` statement prints the final results.
- The condition `D[j-1] != D[j+1]` evaluates to `True` if the nearest neighbors of cell `j` are in different states.
- The line `C[j] = 1 - C[j]` flips the state of cell `j`. If the old state of `C[j]` is 0, the new state is 1. If the old state of `C[j]` is 1, the new state is 0.
- In the `for j in range(1,19):` statement, `j` takes values from 1 through 18. Note that `C = np.zeros(20,dtype=int)`

[1]Roundoff errors do not actually affect this code, because Python can represent the real numbers 0.0 and 1.0 exactly. Nevertheless, it is important to remember that in many situations roundoff error can affect a comparison between real numbers.

creates cells numbered from 0 through 19. We cannot implement our rules for cell number 0 or for cell number 19, since these cells don't have two nearest neighbors. We only implement the rules for the *interior* cells, those numbered 1 through 18.

- The statement `D = np.copy(C)` makes a copy of array `C` and calls it `D`. Why is this necessary? Why don't we use the array `C` throughout the code? Because the element `j` is being modified, or not, depending on the value of element `j-1`. We cannot change the state of element `j-1` before checking to see if `j-1` and `j+1` are in the same state.

The final result for this cellular automaton is

```
[0 0 0 0 0 1 1 1 0 1 1 1 0 1 1 1 0 0 0 0]
```

Exercise 6.8a

Run the cellular automaton code. Modify the code by replacing `D = np.copy(C)` with the simple statement `D = C`. Why doesn't this work?

Remember: The statement `D = C` simply creates another name for the array `C`. When the elements of `C` are modified, the elements of `D` are also modified. On the other hand, `D = np.copy(C)` creates a new array `D` whose elements are initially the same as the elements of `C`. In this case, `C` and `D` can be modified without affecting each another.

How can we visualize the evolution of the cellular automaton? A simple way is to add the following lines of code inside the `for` loop over `n`:[2]

```
for j in range(0,20):
    if C[j] == 1:
        plt.plot(j,n,'r*')
```

You might need to include `plt.close()` at the beginning of your program and `plt.show()` at the end. These lines of code will plot the state of the cellular automaton after each application of the rules, beginning with `n = 0`. You can also display the initial state

[2]Remember, `plt` is shorthand for `matplotlib.pyplot`.

by adding the same lines of code before the loop over `n`, but with `plt.plot(j,n,'r*')` replaced by `plt.plot(j,-1,'r*')`

Exercise 6.8b

Modify the cellular automaton code to produce a plot of the evolution. Increase the number of cells and increase the number of iterations of the rules. Modify the initial state such that several cells are on (+1). See, for example, Fig. (6.1).

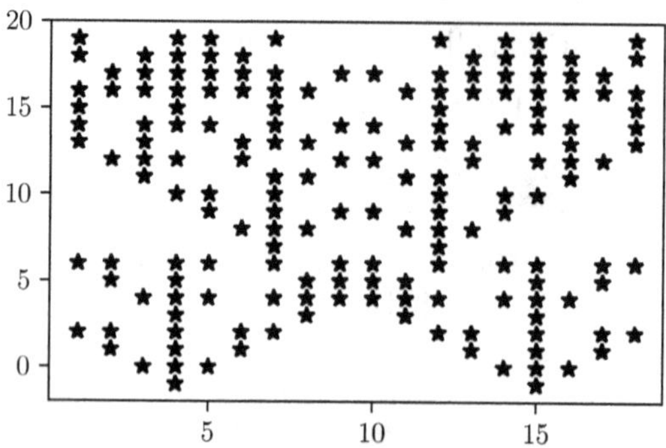

Fig. 6.1. Evolution of the cellular automaton using the rules of Sec. 6.8. Cells 4 and 15 are initially in the on (+1) state. The evolution is carried out for 20 iterations.

Exercise 6.8c

Create a cellular automaton code that uses these rules:

- If cell `j` is off (0), flip it to on (+1).
- If `j` is on and its nearest neighbors `j-1` and `j+1` are the same, leave `j` alone.
- If `j` is on and its nearest neighbors `j-1` and `j+1` are different, flip `j` to off.

Plot the evolution.

Chapter 7

Functions and More Loops

7.1 Function definitions

Common functions such as `sin()` and `sqrt()` are built into the NumPy library. You can define your own functions using the keywords `def` and `return`.

Imagine a code that uses the function $f(x) = \ln(x)\sin(x^3)/(x^2 + xe^x)$ throughout, in many different places. Rather than typing this complicated expression over and over, you can type it once by creating a function definition. The following code defines $f(x)$ with the name `myfxn()`. The main code evaluates `myfxn()` at 3 and prints the result:

```
def myfxn(x):
    return np.log(x)*np.sin(x**3)/(x**2 + x*np.exp(x))

# main code:
y = myfxn(3)
print(f"y = {y:.4}")
```

The output is `y = 0.01517`. This new function can be evaluated throughout the program, as the need arises.

The general form for a function definition is

```
def functionname():
    code to execute
    return value
```

Here,

- *functionname* is the name of the function. The parenthesis `()` contains a list of variables that the function will use.
- *code to execute* is code that is executed as part of the function.
- *value* is the value (or values) that the function returns.

Note that the first line ends with a colon and the remaining lines are indented.

The definition of `myfxn()` given above is particularly simple because there are no lines of "code to execute" between the `def` statement and the `return` statement. Here is another example:

```
def myfxn(x):
    num = np.log(x)*np.sin(x**3)    # compute numerator
    den = x**2 + x*np.exp(x)        # compute denominator
    return num/den

# main code:
y = myfxn(3)
print(f"y = {y:.4}")
```

The "code to execute" can contain loops and control structures, just like any other Python code.

Functions can help streamline your code and make it easier to understand. A function can be defined anywhere, as long as it occurs before the function is used. Most programmers place their function definitions at the beginning of the program, right after the import statements.

Exercise 7.1a

Write a program that defines the function $f(x) = \sin(e^{\sin x})$. Evaluate this function at $x = 0, 1, 2,$ and 3, and print the results.

Exercise 7.1b

Create a function definition for

$$f(x, a) = x^2 + x + a.$$

Have your program compute $f(f(4,2),3)$ and $f(f(4,3),2)$. Which of these numbers is larger? (Try to guess before running your code.)

Exercise 7.1c

Use a Python function to define

$$f(n) = \begin{cases} n/2, & \text{if } n \text{ is even,} \\ 3n+1, & \text{if } n \text{ is odd.} \end{cases}$$

This function acts on a positive integer n to produce another positive integer $f(n)$. Your code should start with an initial choice for n, compute $f(n)$, and iterate; that is, use $f(n)$ as the new value of n, evaluate the function again, and repeat. Have your code print out each integer in the sequence, and stop iterating if the integer reaches 1. The *Collatz conjecture* says that for any initial n, the sequence will always reach 1. What is the longest sequence you can find?

7.2 Functions and variables

Let's take a closer look at functions. Consider a function that computes the length of the hypotenuse of a right triangle using the Pythagorean theorem. The program below defines the hypotenuse function, evaluates it, and prints the result:

```
def hyp(a,b):
    return np.sqrt(a**2 + b**2)


# main code:
c = hyp(3,4)
print("c =", c)
```

The output is `c = 5.0`. Note that `hyp()` is a function of two variables, the lengths of the two short sides of the right triangle.

The complete program consists of two parts, the function definition and the main code. Each part plays a distinct role. The role of the main code:

- When the main code reaches the statement `c = hyp(3,4)`, it passes the values `3` and `4` to the function.
- When the main code receives a value for the hypotenuse from the function, it assigns that value to the variable `c`.

The role of the function:

- When the function receives the values `3` and `4`, it assigns these numbers to the variables `a` and `b`.
- The function computes the hypotenuse `np.sqrt(a**2 + b**2)` and returns the result `5.0` to the main code.

The main code and the function are independent of one another, apart from the passing of numerical values.

The following program is equivalent to the one above:

```
def hyp(a,b):
    return np.sqrt(a**2 + b**2)

a = 3
b = 4
c = hyp(a,b)
print("c =", c)
```

Once again, the main code passes the numbers `3` and `4` to `hyp()`, and `hyp()` returns the result `5.0`.

In the program above, Python does *not* pass the variable names `a` and `b` to the function. It is up to the function to choose variable names for the numbers it receives. The variables `a` and `b` that appear in the function are "local" to the function, and are independent of the variables `a` and `b` that appear in the main code. We could also write this program as

```
def hyp(alpha,beta):
    return np.sqrt(alpha**2 + beta**2)

a = 3
b = 4
c = hyp(a,b)
print("c =", c)
```

The output is the same: `c = 5.0`.

Here is another example:

```
def hyp(a,b):
    return np.sqrt(a**2 + b**2)

a = 17.3
b = 21.8
c = hyp(3,4)
print("c =", c, "a =", a, "b =", b)
```

The function assigns `3` and `4` to its local variables `a` and `b`, computes the hypotenuse, and returns `5.0`. The output of this program is `c = 5.0, a = 17.3, b = 21.8`. The function does not care that the main code has assigned `17.3` and `21.8` to its own variables, also called `a` and `b`. Likewise, the main code does not care that the function has assigned `3` and `4` to its local variables `a` and `b`.

Exercise 7.2a

Write a function that computes the factorial of any positive integer. Use this function in a program that asks the user to input two positive integers, n and m, then computes the number of combinations of n objects taken m at a time:

$$\binom{n}{m} = \frac{n!}{m!(n-m)!}.$$

Test your code to make sure it works correctly.

A function can return more than one number.

Exercise 7.2b

Use a function definition to solve $ax^2 + bx + c = 0$ using the quadratic formula

$$x = \frac{-b \pm \sqrt{b^2 - 4ac}}{2a}.$$

The inputs to the function will be the coefficients `a`, `b`, and `c`. Since there are two solutions, there are two return values. Let's call these values `x1` and `x2`, and name the function `quad()`. Then the return statement is `return x1, x2`. The main-code statement `x1, x2 = quad(a,b,c)` will assign the return values to `x1` and `x2`.

A function can accept a NumPy array as input, as long as the *code to execute* can evaluate the array.

Exercise 7.2c

Create a Python program that defines the function $\sin(a/x)$. Have your code produce a graph of $\sin(3/x)$ for $0.1 \leq x \leq 2.0$. The inputs to the function will be $a = 3$ and a NumPy array of x values. The function will return a NumPy array of $\sin(a/x)$ values.

7.3 Variables and parameters

In the previous section we saw that the variables used in the main code and the variables used in a function definition are isolated from one another. Only numerical values are passed between the main code and the function.

This isn't the full story. If a function depends on a variable whose value has *not* been passed from the main code, the function will "look outside" to the main code for the missing value.

Here is an example:

```
def hyp(a):
    return np.sqrt(a**2 + b**2)

a = 3
b = 4
c = hyp(a)
print("c =", c)
```

The main code passes the value 3 to the function `hyp()`. The function assigns this value to its local variable `a`. The next task for the function is to compute `np.sqrt(a**2 + b**2)`. Since the function doesn't have a value for `b`, it "looks outside" to the main code. It sees that `b` equals 4 in the main code, and uses this value to calculate `np.sqrt(a**2 + b**2)`.

You could also exclude `a` from the argument list of the hypotenuse function. It is probably not a good idea to exclude either `a` or `b`, or both. As a general rule, all of the variables used in a function definition should be included in the argument list. This makes the function easier to understand and easier to reuse in another code.

Let's look at another example. A ball is thrown straight up into the air. The ball's height y as a function of time t is

$$y = y_0 + v_0 t - \frac{1}{2} g t^2, \tag{7.1}$$

where y_0 is the initial height, v_0 is the initial velocity, and g is the acceleration due to gravity. The program below uses a function definition to compute the ball's height at various times t:

```
def height(y0,v0,g,t):
    return y0 + v0*t - 0.5*g*t**2

y0 = 2.0        # initial height (m)
v0 = 10.0       # initial velocity (m/s)
g = 9.8         # acceleration (m/s^2)

for t in [0.0, 0.4, 0.8, 1.2, 1.6, 2.0]:
    y = height(y0,v0,g,t)
    print(f"y({t:.1f}) = {y:.3f}")
```

Exercise 7.3a

Run the code above. What is the output?

The function `height()` depends on four variables, y_0, v_0, g, and t. The values of these variables are passed from the main code to the function, which returns the height of the ball.

Using terminology that is common in physics and mathematics, we might describe y_0, v_0, and g as *parameters* and t as a *variable.* The difference is this: Parameters have fixed values throughout the program, whereas variables take on multiple values. This is a somewhat loose distinction, but is usually clear in practice. In our "ball in air" program, the parameters y_0, v_0, and g never change, but the variable t takes on multiple values as the program unfolds.

If a function depends on many parameters, it can be tedious to include all of the parameters in the argument list. As a practical matter, it might be preferable to drop the parameters from the argument list. For example, we can drop `y0`, `v0`, and `g` from the list of arguments for the height function:

```
def height(t):
    return y0 + v0*t - 0.5*g*t**2

y0 = 2.0        # initial height (m)
v0 = 10.0       # initial velocity (m/s)
g = 9.8         # acceleration (m/s^2)

for t in [0.0, 0.4, 0.8, 1.2, 1.6, 2.0]:
    y = height(t)
    print(f"y({t:.1f}) = {y:.3f}")
```

Only the variable `t` is explicitly passed to `height()`. The function looks to the main code for values of the parameters `y0`, `v0`, and `g`.

Exercise 7.3b

Verify that the program produces the correct output when the parameters are omitted from the argument of `height()`.

From the computer's point of view, there is really no distinction between parameters and variables. Thus, we could modify our code even further by dropping t from the argument list. That's probably not a good idea. As a general rule, all variables used in a function should be included in the argument list. Ideally, parameters should be included as well. However, this might be impractical if the number of parameters is large.

Exercise 7.3c

Newton's universal law of gravitation says that the gravitational force between objects of mass m_1 and m_2 is

$$F = \frac{G m_1 m_2}{r^2},$$

where $G = 6.67 \times 10^{-11}\,\mathrm{N\,m^2/kg^2}$ is Newton's constant. Here, r is the distance between the two objects. Write a program to compute the force between Earth ($m_2 = 5.97 \times 10^{24}$ kg) and the moon ($m_1 = 7.35 \times 10^{22}$ kg) at various distances. Use a function definition for F, treating r as a variable and G, m_1, and m_2 as parameters. The distance between Earth and the moon varies from roughly $r = 3.6 \times 10^8$ m to $r = 4.1 \times 10^8$ m. How much does the force vary?

7.4 Beyond functions

In mathematics a function is a "machine" that takes an input (or inputs) and produces an output (or outputs). For most applications in scientific computing, we can think of Python functions in the same way. For example, the definition

```
def F(a,x):
    v = a*x
    return v
```

is the programming equivalent of $F(a, x) = ax$. The inputs are a and x, the output is ax.

A Python function definition can go beyond the usual mathematical concept of a function by modifying the inputs. This can be a source of error if not done carefully. Consider the main code

```
a = 2.0
x = np.array([1.1, 2.2, 3.3])
y = F(a,x)
print(f"a = {a}, x = {x}, y = {y}")
```

In the first line, `a` is given the value `2.0` and in the second line, `x` is assigned to the array `[1.1 2.2 3.3]`. In the third line, the function `F()` is used to multiply each element of `x` by `a` and return the result. This program produces the output

```
a = 2.0, x = [1.1  2.2  3.3], y = [2.2  4.4  6.6]
```

Exercise 7.4a

Make the following modifications, one at a time, to the program above.

- Add the statement `a = 3.0` inside the function definition, before `v = a*x`. Does this change the value of `a` in the main code? Does this change the array `y`?
- Add the statement `x = np.array([2.2, 3.3])` inside the function definition, before `v = a*x`. Does this change `x` in the main code? Does this change `y`?
- Add the statement `x[1] = 4.4` inside the function definition, before `v = a*x`. Does this change `x` in the main code? Does this change `y`?

The statements `a = 3.0` and `x = np.array([2.2,3.3])` inside the function definition do not affect the values of `a` or `x` outside the function, but they do affect the result for `y`. On the other hand, the statement `x[1] = 4.4` inside the function definition changes the array `x` both inside and outside the function.

This leads us to another general rule: The return statement should include all variables and parameters whose values might be changed by the function. This is the best way to insure that changes made inside the function are passed back to the main code. In the example above, we should include `a` and `x` in the return statement:

```
def F(a,x):
    statements that modify a and x
    y = a*x
    return y, a, x
a = 2.0
x = np.array([1.1, 2.2, 3.3])
y, a, x = F(a,x)
print(f"a = {a}, x = {x}, y = {y}")
```

Exercise 7.4b

Verify that this code works as expected by placing statements such as `a = 3.0`, `x = np.array([2.2, 3.3])` and `x[1] = 4.4` inside the function definition.

7.5 More `for` loops

In Sec. 5.5, we considered a code that computes the square roots of the numbers from 0 through 10 in steps of 0.5. The code below produces the same result with a more readable output:

```
A = np.linspace(0,10,21)
for i in range(len(A)):
    print(f"The sqrt of {A[i]:.2f} is {np.sqrt(A[i]):.4f}")
```

Here, we are using a `for` loop to create a separate print statement for each element of `A`. As a byproduct, we can compute the square root values "on the fly" inside the `print` command.

In the code above the `for` loop cycles over values of the array index. The following code achieves the same result by letting the `for` loop cycle over the values of the array itself:

```
A = np.linspace(0,10,21)
for x in A:
    print(f"The sqrt of {x:.2f} is {np.sqrt(x):.4f}")
```

The command `for x in A` tells Python to repeat the indented code for each element `x` in the array `A`. In the first pass through the loop, `x` is set equal to `A[0]`; in the second pass through the loop, `x` is set equal to `A[1]`; etc.

Exercise 7.5a

Create an array such as `A = np.array([-2.3, 5.4, 7.1, -4.2, 6.8])`. Write a code that will compute the sum of the array elements. Do this in two ways, by looping over the array index and by looping over the array elements. (The NumPy function `sum(A)` will sum the elements of an array `A`. Use this to check your answers.)

The following code computes the exponentials of the numbers 0.0 through 2.0 in increments of 0.1:

```
for n in range(0,21):
    x = n/10
    expx = np.exp(x)
    print(f"exp({x:.1f}) = {expx:.3f}")
```

Here, we created the numbers $x = 0.0, 0.1, \ldots, 2.0$ by dividing the integers $n = 0, 1, \ldots, 20$ by 10.

Exercise 7.5b

Create the numbers 0.0 through 2.0 in increments of 0.1 using the NumPy `linspace` command. Use a `for` loop to cycle through those numbers. Compute and print the exponential of each number.

Exercise 7.5c

Write a code that accepts a word as user input and counts the number of vowels in the word. (Vowels are the letters "a", "e", "i", "o", and "u".)

7.6 `while` loops

A `for` loop is useful when we know (or the code can compute ahead of time) the number of times the loop should be executed. A `while` loop is useful when we don't know how many times the loop should be executed. Rather, the number of iterations of the loop depends on the results of calculations within the loop.

The syntax for the `while` loop is

> `while` *condition*:
>
> *code to execute and repeat as long as condition is* `True`

The code to be executed must be indented.

The following code computes the exponentials of $0.0, 0.1, \ldots$ etc. and continues until the result exceeds 1000:

```
x = 0.0
expx = 0.0
while expx < 1000:
    expx = np.exp(x)
    print(f"exp({x:.1f}) = {expx:.3f}")
    x = x + 0.1
```

When Python reaches the `while` statement, it executes the loop repeatedly, as long as the condition `expx < 1000` evaluates to `True`. The program exits the loop when the condition evaluates to `False`.

Some things to note about this code:

- The initial value of `x` is placed before the `while` loop, and `x` is incremented by 0.1 for each pass through the loop.
- The variable `expx` is given a value 0.0 before the `while` loop. This allows Python to evaluate the conditional `expx < 1000` when it first reaches the `while` statement. The initial value for `expx` doesn't matter, as long as it is less than 1000.

Exercise 7.6a

The code above doesn't stop looping until `expx` exceeds 1000. As a result, in the final line from the `print` command, `expx` is greater than 1000. This might not be what we intended. Rewrite the code so that it only prints the results with `expx < 1000`.

Recall that conditions are statements that evaluate to `True` or `False`. Most conditions involve a comparison between two quantities using a comparison operator, such as == or <. Python also defines certain quantities as intrinsically `True` or `False`. For example, nonzero numbers and the word `True` will evaluate to `True`. The number zero and the word `False` will evaluate to `False`.

It is common to use the conditional `True` with a `while` loop, along with the command `break` to exit from the loop. As an example, let's find the cubes of all positive integers until the result exceeds some maximum value, say, 2000. Here's one way:

```
n = 1
while True:
    x = n**3
    if x <= 2000:
        print(x)
        n = n + 1
    else:
        break
```

Comments:

- The counter `n` is initialized to `1` before the loop begins.
- The condition `True` always evaluates to `True`.
- The `if` statement checks to see if `x` is less than or equal to `2000`.
- If `x <= 2000` then `x` is printed, the counter is incremented, and the loop is repeated.
- If `x` is not `<= 2000` (that is, `x` is greater than `2000`) then the `break` command causes the `while` loop to terminate.

Exercise 7.6b

Write a code to find the largest sum of integers $1+2+3+\cdots$ that is less than 10^6 (one million).

Exercise 7.6c

Create an array containing all of the numbers of the form 3^n (where n is a positive integer) that are less than 10000.

Warning: Don't use `while True` without `break`. Since `True` always evaluates to `True`, the `while` loop will continue running forever, in an "infinite loop," if there is no `break` command.

What should you do if your code is stuck in an infinite loop? You can always stop the program while it is running. Look for a button or a drop-down menu item that says something like "interrupt execution" or "interrupt kernel."

7.7 More exercises

Exercise 7.7a

The *floor function* is defined as the largest integer less than or equal to its input. As a function of x, the floor function is often denoted $\lfloor x \rfloor$. For example, $\lfloor 3.26 \rfloor = 3$ and $\lfloor -4.6 \rfloor = -5$. NumPy has a built–in floor function called `floor()`. Create your own floor function and test it using various real numbers, both positive and negative.

Exercise 7.7b

The *Chebyshev polynomials* of the first kind are defined by

$$T_n(x) = \cos(n \arccos(x)),$$

where n is a nonnegative integer and $-1 \leq x \leq 1$. (You can use multiple–angle trig identities to show that $T_n(x)$ is an nth order polynomial in x.) Create a function definition for the Chebyschev polynomials and plot T_0 through T_5 on the same graph.

Chapter 8

Random Topics

8.1 Subplots, legends and Mathtext

The basic `plot()` function in Matplotlib is adequate for many purposes. The `subplots()` function opens up a wider range of functionality.

Consider a mass on a spring. The mass executes simple harmonic motion $x(t) = A\sin(\omega t)$ with amplitude A and angular frequency ω. After importing NumPy as `np` and `matplotlib.pyplot` as `plt`, the following code produces two graphs, one for position $x(t)$ and the other for velocity $v(t) = dx/dt = A\omega\cos(\omega t)$:

```
# Choose parameter values
A = 2.0
omega = 5.0

# Create arrays
t = np.linspace(0,5,100)
x = A*np.sin(omega*t)
v = A*omega*np.cos(omega*t)

# Create plots
plt.close()
fig, (ax1, ax2) = plt.subplots(2,1)

ax1.plot(t,x)
ax1.set_xlabel('t')
ax1.set_ylabel('x(t)')
```

```
ax2.plot(t,v)
ax2.set_xlabel('t')
ax2.set_ylabel('v(t)')
plt.show()
```

Exercise 8.1a

Run this code and observe the output. What happens if you replace `subplots(2,1)` with `subplots(1,2)`?

The command

```
fig, (ax1, ax2) = plt.subplots(2,1)
```

instructs Python to create a figure named `fig`. The figure consists of two plots, referred to as *axes*. The axes (plots) are named `ax1` and `ax2`. The argument `(2,1)` says that the plots should be arranged into 2 rows and 1 column.

The command `ax1.plot(t,x)` plots x versus t on the first axis. The command `ax2.plot(t,v)` plots v versus t on the second axis. The plot labels are created with `set_xlabel()` and `set_ylabel()`.

You can specify the size of the figure by adding the option `figsize` as an argument to `subplots()`. For example,

```
fig, (ax1, ax2) = plt.subplots(2,1,figsize=(5,7))
```

creates a figure that is 5 inches wide and 7 inches high. The default values are 6.4 in × 4.8 in.

To add figure and plot titles, insert the commands

```
fig.suptitle('Simple Harmonic Motion')
ax1.set_title('position')
ax2.set_title('velocity')
```

somewhere between `plt.close()` and `plt.show()`. You might need to add `plt.tight_layout()` before `plt.show()` to insure proper spacing between plots.

Exercise 8.1b

Add figure and plot titles, and experiment with different figure sizes. How does `plt.tight_layout()` affect your figure?

When there are multiple curves on a single plot, it can be helpful to add a legend. Here is a code that plots functions of the form $ax\sin(x)$ for several values of a, and includes a legend:

```
# Define functions
def f(x,a):
    return a*x*np.sin(x)

# Choose parameter values
a1 = 2.0
a2 = 3.0
a3 = 4.0

# Create x array
x = np.linspace(0,10,200)

# Create figure
plt.close()
fig, (ax1) = plt.subplots(1,1)
ax1.plot(x,f(x,a1),label='a ='+str(a1))
ax1.plot(x,f(x,a2),label='a ='+str(a2))
ax1.plot(x,f(x,a3),label='a ='+str(a3))
ax1.legend()
plt.show()
```

In this example we have chosen to use **subplots()**, even though there is only one plot. This is a common practice.

Exercise 8.1c

Plot functions of the form $x\cos(ax)$ on a single figure using various values of a. Include a legend.

By default, Matplotlib will display each curve as a solid line and cycle through the colors in this order: blue, orange, green, red, purple, brown, pink, gray, olive, and cyan. You can override these colors by adding a color option to the plot command. The default colors are specified by 'C0' for blue, 'C1' for orange, etc. For example,

```
ax1.plot(x,f(x,a1),'C2', label='a ='+str(a1))
```

will set the color of the first curve (the curve with parameter value a1) to green. Colors can also be specified as 'b' (blue), 'g' (green), 'r' (red), 'c' (cyan), 'm' (magenta), 'y' (yellow), 'k' (black), or `'w'` (white). Note that `'b'` is not the same shade of blue as `'C0'`; likewise for the other colors.

The linestyle is specified with options such as `'-'` (solid), `'--'` (dashed), and `':'` (dotted). The option `'.'` tells Matplotlib to plot only the data points. The linestyle should be combined with the color option, if both are specified. For example,

```
ax1.plot(x,f(x,a1),'b:', label='a ='+str(a1))
```

creates a blue, dotted curve.

Exercise 8.1d

Using the code from the previous exercise, experiment with different colors and line styles. How do you make a "dash-dot" curve?

One final tip: You can use *Mathtext* in titles, labels and legends. Mathtext is a subset of TeX, used for typesetting mathematical expressions. Strings encompassed by dollar signs `$` are processed as Mathtext. For example,

```
fig.suptitle(r'$\frac{d^2 x}{dt^2} = -\omega^2 x$')
```

creates the title $\frac{d^2x}{dt^2} = -\omega^2 x$. The `r` in front of the string tells Python that what follows is a "raw string."

8.2 Data input and output

Computers are often used to handle large amounts of data. Computer programs should be capable of reading data from a text file, and writing data to a text file. For scientific purposes, we usually want to read and write arrays of numbers. NumPy includes the functions `savetxt()` and `loadtxt()` that make reading and writing data straightforward.

To save a one-dimensional array `A` to a file named *filename.txt*, use

```
np.savetxt("filename.txt",A)
```

To read data from a file named *filename.txt* into an array `A`, use

```
A = np.loadtxt("filename.txt")
```

Remember, the file name is a string and should be surrounded by quotation marks.

If you are running Python on your local machine, the program will generally use your local working directory for reading and saving files. That is, the directory where the program itself is saved. If you are running Python in the cloud, reading and writing files to and from your local computer might require some extra steps.

Exercise 8.2a

Run the code

```
A = np.linspace(0,5,20)
np.savetxt("myarray.txt", A)
```

Does the file *myarray.txt* appear in the expected location? Does it have the right content? Now read the file *myarray.txt* into a NumPy array `B`:

```
B = np.loadtxt("myarray.txt")
print(B)
```

Does `B` equal `A`?

For a one-dimensional array, the numbers in a file can be written as either a single row or a single column.

Exercise 8.2b

Using a text editor, create a text file with several numbers in a single column. Create a second text file with the same numbers listed in a row, separated by spaces. Use `loadtxt()` to load these files into NumPy arrays `A` and `B`. Does `A = B`? What happens if you use commas to separate the numbers in the second text file?

The `savetxt()` and `loadtxt()` functions can be used with multicolumn data.

Exercise 8.2c

Create a text file consisting of two columns, each column containing several numbers. Use `loadtxt()` to read the file into an array `C`. What does `print(C)` produce?

In the exercise above, `loadtxt()` reads the data into a two-dimensional NumPy array `C`. Recall that you can reference individual elements of a two-dimensional NumPy array using the syntax `C[i][j]` or, more simply, `C[i,j]`.

If you want each column of the file to occupy a single one-dimensional array, you can copy the two columns of `C` into `A` and `B` using

```
A = C[:,0]      # Copy first column of C into A
B = C[:,1]      # Copy second column of C into B
```

The colon : means "all elements."

Exercise 8.2d

Examine the array `C` from the previous exercise. Which element is `C[0,1]`? Which element is `C[1,0]`? Place each column of `C` into a separate array. Place each row of `C` into a separate array.

There is a more direct way to load a multi-column file into separate one-dimensional arrays:

```
a,b = np.loadtxt("filename.txt", unpack=True)
```

This command extracts the first column of *filename.txt* (which has index `0`) and places it into the one-dimensional array `a`. The second column (which has index `1`) is placed into the one-dimensional array `b`.

Exercise 8.2e

Consider once again the multi-column file from the previous exercises. Read the columns of this file directly into one-dimensional arrays and verify the results.

The `savetxt()` function will place the elements of a one-dimensional array into a file with a single column. We can save multiple one-dimensional arrays into the columns of a single file. For example, let's say our code computes the height of a ball as a function of time. Let `t` denote the array of times and `y` denote the array of heights. The command

```
np.savetxt("filename.txt", list(zip(t,y)))
```

will create the file *filename.txt* with `t` in the first column and `y` in the second column.

Exercise 8.2f

Use Eq. (7.1) to compute the height y of a ball as a function of an array of times t. Have your code save the data to a single file with t in the first column and y in the second column.

Exercise 8.2g

Write a program that uses **def** to define the function $y = e^{\sin(x)}$. Have your code compute y for at least 100 values of x from $-\pi$ to π. Output the x and y data to a file. Write a second program that will read the data from the file and plot a graph of y versus x.

8.3 Random numbers

In scientific programming we sometimes need random numbers. How do we instruct the computer to choose a random number? The NumPy library includes several random number generators. We will use the default generator. Start with the command

```
rnum = np.random.default_rng()
```

This tells Python to prepare a random number generator and call it `rnum`. We can now issue the command

```
rnum.integers(low,high)
```

which yields a random integer from *low* to *high* (including *low*, excluding *high*). Each time this command is used in a program, the random number generator creates a new random integer.

Exercise 8.3a

Write a code that repeatedly produces and prints random integers between *low* = 1 and *high* = 20. Design your code to stop if the random number equals 5.

The command

```
rnum.uniform(low,high)
```

yields a random real number from *low* to *high* (including *low*, excluding *high*). Each number in the range is equally likely; hence the name `uniform`. Every time this command is used in a program, the random number generator creates a new random number.

Exercise 8.3b

Create a code that produces N random real numbers from *low* = 0 to *high* = 1. Find the average of these numbers. Does the average tend to converge to 0.5 as N increases from 10^2 to 10^4? 10^6?

Random numbers generated by a computer aren't really random. After all, a computer can only follow a prescribed algorithm, and algorithms are fundamentally deterministic. The numbers generated by the NumPy (or any other) random number generator are more accurately characterized as *pseudorandom*. For many purposes, they are "random enough."

If the computer follows a deterministic algorithm, how does it produce different random numbers each time you run the code? The answer is that the algorithm itself depends on a "seed" number. By default, the seed number is obtained from the operating system using various hardware sources such as mouse movements and fan noise. Instead of using the default seed, we can choose a seed number explicitly,

```
rnum = np.random.default_rng(seednum)
```

Note that *seednum* must be an integer. When developing a code that uses random numbers, it can be useful to assign a seed number explicitly. This causes the results to be reproducible.

Exercise 8.3c

Specify a seed number in your code from Exercise 8.3a. Run the code multiple times to make sure the results are reproducible. Try different seed numbers.

8.4 Real number formats

In Sec. 2.9 we introduced f-string formatting. Given a variable `x = 23.39128287` the print statement `print(f"x = {x:5f}")` produces the output `x = 23.39128`. The `f` before the opening quotation mark tells Python to interpret the string as an f-string.

The number `23.39128` that appears in the output is produced by the part of the string between curly braces, namely, `x:.5f`. A colon separates the number to be printed, `x`, and the format of the number, `.5f`. This format tells Python to print `x` as a real number (float) rounded to `5` decimal places. That is, with `5` digits after the decimal point.

There are other formats. For example, `x:.5e` specifies scientific notation (exponential notation) with `5` digits after the decimal point. The "general" format `x:.5g` will print `x` to `5` significant figures. Examples of these formatting types are given in Table 8.1. The underbracket shows the digits after the decimal point (for `f` and `e` formats), or the significant figures (for `g` format).

These format types can be used with the NumPy `savetxt()` function as well. For example, `np.savetxt("`*filename.txt*`",A,"%.5f")`

Table 8.1. Real numbers `x` from the first column are shown with `.5f` (float), `.5e` (exponential) and `.5g` (general) formatting in the remaining columns. The digits specified by the number `5` are displayed with an underbracket.

x	x:.5f	x:.5e	x:.5g
23.39128287	$23.\underbrace{39128}$	$2.\underbrace{33913}\text{e}+01$	$\underbrace{23.391}$
0.003248124	$0.\underbrace{00325}$	$3.\underbrace{24812}\text{e}-03$	$0.00\underbrace{32481}$
7424681.198	$7424681.\underbrace{19800}$	$7.\underbrace{42468}\text{e}+06$	$\underbrace{7.4247}\text{e}+06$

will save the contents of the numerical array `A`, with the numbers rounded to `5` decimal places.

> **Exercise 8.4**
>
> Create a NumPy array of random real numbers from 0 to 10. Save the array to a text file with the numbers expressed to 8 significant figures.

8.5 Negative indexing

Let `A` be either a list or an array with `N` elements. The first element is `A[0]`, the second element is `A[1]`, etc. The last element is `A[N-1]`. Each of the elements can also be accessed with *negative indexing.* That is, for $i = 0, 1, \ldots, N-1$, element `A[i]` is also denoted `A[i-N]`. The first element of `A` is `A[0]` and `A[-N]`. The second element is `A[1]` and `A[1-N]`. The last element is `A[N-1]` and `A[-1]`. This is depicted in Fig. 8.1.

Negative indexing can be useful in various contexts. For example, we might want to access the final element of an array `A`, but we don't immediately know the size of the array. The last element is always `A[-1]`.

> **Exercise 8.5**
>
> Create an array of numbers of the form n^3 that are less than 2500, where n is a positive integer. After the array has been created, have your code print out the largest element.

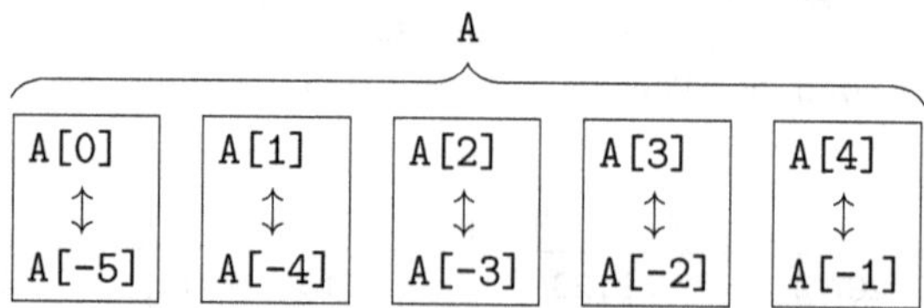

Fig. 8.1. A list or array `A` with $N = 5$ elements. The individual elements are accessed with positive indices 0 through 4, as well as negative indices -1 through -5.

8.6 Object oriented programming

Classes, objects and methods form the foundation of "object oriented programming" (OOP). You don't need to be an expert in object oriented programming to understand the principles of scientific computation. But a basic understanding of methods, in particular, will help explain the syntax that we sometimes encounter.

An *object* is a member of a *class*. A *method* is a function that acts on an *object*. Let's look at a very simple example.

A regular polygon is a multi-sided figure with all sides of equal length, and equal interior angles. The code below creates the class of regular polygons. Objects in this class are triangles, squares, pentagons, etc. of various sizes. Each object has the following attributes: a name (called **name**), a certain number of sides (called **nsides**), and a particular side length (called **lsides**). Within the polygon class we define a *method* called **area()**, which is a function that computes the area of a polygon object. Here is the code:

```
class polygon:
    def __init__(self,name,nsides,lsides):
        self.name = name
        self.nsides = nsides
        self.lsides = lsides
    def area(self):
        N = self.nsides
        L = self.lsides
        return N*L**2/(4*np.tan(np.pi/L))
```

We can create a "big triangle" object with

```
BT = polygon('big triangle',3,12)
```

and a "small square" object with

```
SS = polygon('small square',4,1/3)
```

Now apply the **area()** method to find the area of each object. We do this by appending **.area()** to the end of the object. Thus,

```
print(f"area of {BT.name} is {BT.area():.2f}")
print(f"area of {SS.name} is {SS.area():.3f}")
```

produces the output

```
area of big triangle is 62.35
area of small square is 0.111
```

Exercise 8.6a

Add a method to the polygon class to compute the circumradius of a polygon object. (The circumradius $R = L/(2\sin(\pi/N))$ is the distance from the center to one of the vertices.) Now create an octagon with sides of length 2. What are the area and circumradius of this octagon?

We previously encountered methods in Sec. 8.3 on random numbers. The command

```
rnum = np.random.default_rng()
```

creates a random number object called `rnum`. The methods `integers()` and `uniform()` act on the object to produce random integers and random real numbers.

Exercise 8.6b

In Sec. 2.8, we introduced the method `split()`, which converts a string into a list of strings. Create a string `myname` consisting of your name. Compare `print(myname)` to `print(myname.split())`. What data type is `myname`? What data type is `myname.split()`?

Chapter 9

Programming Practice

9.1 Building complex programs

You should build your programs using small steps that you can test along the way. Consider a program that simulates a simple guessing game. The user is asked to guess an integer from 1 through 1000, and the code responds with “too high” or “too low” or “correct!”. The user continues to guess until they get the right answer.

Exercise 9.1

Create this program in small steps, following the outline below.

To begin, the code needs to pick a random number from 1 through 1000. After importing NumPy and creating a random number generator called `rnum`, we write

```
thenum = rnum.integers(1,1000)
```

Is this statement correct? Should we use `rnum.integers(1,1001)` instead? Or perhaps `rnum.integers(0,1000)`? Try a quick experiment.

Replace `rnum.integers(1,1000)` with `rnum.integers(1,5)` and add a `print(thenum)` statement. Run the code repeatedly. Which numbers appear?

This experiment reminds us that `rnum.integers(`*low*`,`*high*`)` creates integers from *low* to *high* with *low* included but *high* excluded. If we want random numbers chosen from the set $1, 2, \ldots, 1000$, we need to use `rnum.integers(1,1001)`.

> Correct the `rnum.integers()` function and remove your print statement.

The next step in building the code is to prompt the user for a guess. We can do this with the `input()` function, something like

```
guess = input("guess a number from 1 through 1000: ")
```

> Add this to your code and run it.

Now add some code to check whether the guess is too high or too low. We can worry later about the case in which the guess is exactly correct. Extend your code by adding the lines

```
if guess > thenum:
    print("too low")
else:
    print("too high")
```

> Run the code. Does it work?

You probably received a "TypeError" and a cryptic message about the use of $>$ with `int` and `str`. After some thought, we recall that the `input()` function treats user input as a string.

> Convert the input to an integer by adding the command `guess = int(guess)`. Does your code work now? It's probably difficult to tell. Add a print statement to print out `thenum`, so you can verify if your guess is too high or too low. Run the code several times to see if it works as expected.

Oops! When `guess` is less than `thenum`, the code prints "too high." When `guess` is greater than `thenum`, the code prints "too low."

> Modify your code to correct this error.

Your code should work correctly so far. But it doesn't check to see if `guess` is correct.

> Use the `if`, `elif`, `else` construction to extend your code to account for the possibility that `guess` is correct.

How can we test this when we only have a 1 in 1000 chance of getting the right answer? If you still have a print statement that tells the value of `thenum`, you can cheat and input the correct answer.

> Test your code. It should give the right response whether `guess` is too high, too low, or correct. Remember, you can always specify a seed number for the random number generator to get reproducible results.

Finally, we need to modify the code to allow the user to make multiple guesses. If the number of guesses is unlimited, we can use a `while` loop that is always `True`, then `break` from the loop when `guess` is correct.

> Add the statement `while True:` before the input. Indent all of the code that should be repeated for multiple guesses. You can instruct your code to break out of the `while` loop by adding the command `break` after the `print("correct!")` statement. Test your code!

When you're satisfied that the code is working correctly, you should remove (or comment out) the print statement that gives away the answer. Also delete the seed number, if you have one.

Remember—always build your program in small logical steps, testing each new part along the way. It is much easier to catch mistakes and logical errors with this approach. Start practicing these good habits now, while your codes are relatively simple, and you will develop the skills to create more complex, error-free programs.

9.2 Roulette

Roulette is a casino game that consists of a wheel whose outer ring is divided into 38 slots. The wheel is spun in one direction and a small ball is rolled in the opposite direction. When the wheel comes to rest the ball falls randomly into one of the 38 slots. The slots on the roulette wheel are numbered from 1 through 36, plus 0 and 00.[1]

There are many different ways a player can place a bet. For our purposes we will consider one of the most simple options: The player bets that the ball will land on one of the positive even numbers $2, 4, 6, \ldots, 36$. This type of bet has a one-to-one payout. In other words, with a \$5 bet, the player will win \$5 if the ball lands on an even number, and lose \$5 if the ball lands on either an odd number or 0 or 00. The odds of winning a single bet of this type are 18/38, slightly less than 50%.

The martingale betting system is a scheme that appears to allow the player to "beat the odds." Here's the idea. The player plays repeated rounds, starting with an initial bet of, say, \$1. Every time the player loses, they double their bet and play again. Every time the player wins, they start over with the initial \$1 bet.

It seems like the martingale system will work as long as the player ends their gambling session with a win. For example, let's start with \$20. A gambling session might look like this:

round	bet	result	gain/loss	total
1	\$1	loss	−\$1	\$19
2	\$2	win	+\$2	\$21
3	\$1	loss	−\$1	\$20
4	\$2	loss	−\$2	\$18
5	\$4	win	+\$4	\$22
6	\$1	loss	−\$1	\$21
7	\$2	win	+\$2	\$23

With each win, the player recovers any previous losses and gains another dollar in the process.

[1]Some roulette wheels have 37 slots numbered 0 through 36.

Before you empty your bank account and rush off to the casino, you might want to test the martingale betting system with some computer simulations.

Exercise 9.2

Write a code to simulate a player who plays roulette, always bets on "even," and uses the martingale betting system. The minimum bet is $1 and the player has $20 to gamble. The output of your code might look something like this:

```
round 1 bet is $1
ball lands on 23
you lose $1, you have $19

round 2 bet is $2
ball lands on 16
you win $2, you have $21

round 3 bet is $1
ball lands on 0
you lose $1, you have $20

round 4 bet is $2
ball lands on 7
you lose $2, you have $18

round 5 bet is $4
ball lands on 34
you win $4, you have $22

round 6 bet is $1
ball lands on 11
you lose $1, you have $21

round 7 bet is $2
ball lands on 28
you win $2, you have $23
```

The player can double their bet after each loss only as long as they have enough money to do so. If not, the player just bets as much as possible.

The play must stop if the player is bankrupt. Let's also assume that the player is not greedy; their goal is come away with \$100. Write your code such that the play stops if either:

- The player's money reaches \$100. (The code should print out a message such as "Congratulations!".)
- The player's money reaches \$0. (The code should print out a message such as "Sorry, you're bankrupt".)

Run your code many times. How often are you bankrupt? How often do you win \$100?

The martingale system fails because it assumes the player has an infinite amount of money to gamble. In reality, the player will always, eventually, hit a string of losses that is long enough to bankrupt them. Even if the player has a near-infinite amount of money, the casinos always limit the size of each bet. So for a sufficiently long string of losses, the player will not be able to double their bet. More often than not, the gambler who uses the martingale strategy will lose money.

9.3 More cellular automata

As described in Sec. 6.8, a cellular automaton consists of a one-dimensional grid of cells. The state of each cell is either on $(+1)$ or off (0), and the cells evolve according to a given set of rules. At each stage of the evolution, the state of cell j is modified depending on the states of its nearest neighbors, cell $j-1$ on the left and cell $j+1$ on the right. The rules can't be applied to the first cell, since the first cell has no nearest neighbor to the left. The rules can't be applied to the last cell, since the last cell has no nearest neighbor to the right.

Imagine bending the grid of cells into a circle, so that the first and last cells are next to one another. This forms a cellular automaton with "periodic boundary conditions." Every cell has two nearest neighbors, and the evolution rules can be applied to all cells.

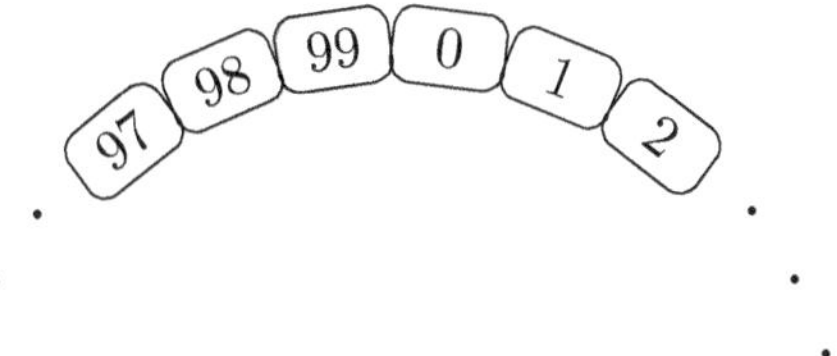

As an example, suppose you choose 100 cells, numbered from 0 to 99. As usual, the nearest neighbors of cell 27 are cells 26 and 28. With periodic boundary conditions, cell 0 has nearest neighbors 99 and 1. Cell 99 has nearest neighbors 98 and 0. The evolution rules can be applied to all cells, 0 through 99.

Exercise 9.3a

Create a cellular automaton code with periodic boundary conditions. Use the rules of Sec. 6.8 and experiment with different initial states.

Exercise 9.3b

Create a cellular automaton code with periodic boundary conditions that applies "Rule 30," which says:

- If cells `j` and `j-1` are both on (+1), flip cell `j` to off (0).
- If cell `j` is on and cell `j-1` is off, leave `j` on.
- If cell `j` is off and cells `j-1` and `j+1` are different, switch `j` to on.
- If cell `j` is off and cells `j-1` and `j+1` are the same, leave `j` off.

Exercise 9.3c

Simulate a cellular automaton (with or without periodic boundary conditions) using your own rules.

9.4 Mandelbrot set

The Mandelbrot set is named after Benoit Mandelbrot who carried out pioneering work on fractals in the early 1980's.[2] The set was described in earlier works by Brooks and Matelski.[3]

The Mandelbrot set is defined in terms of a function $F(z)$ on the complex plane. The function is described in terms of the sequence

$$u_{n+1} = (u_n)^2 + z, \tag{9.1}$$

with $u_0 = 0$. Here, $(u_n)^2$ is the square of the complex number u_n. For example, with $z = -1.0 + 0.5i$, the sequence is $u_0 = 0.0$, $u_1 = -1.0 + 0.5i$, $u_2 = -0.25 - 0.5i$, etc. Results are collected in Table 9.1 along with $|u_n|$, which is the absolute value of u_n. Now ask: How many iterations of Eq. (9.1) are required for $|u_n|$ to exceed 2? Table 9.1 shows that $|u_5| > 2$, so five iterations are required when $z = -1.0 + 0.5i$.

The number of iterations required for $|u_n|$ to exceed 2 depends on z. Once $|u_n|$ exceeds 2, it quickly diverges to infinity. However, for some values of z, $|u_n|$ never exceeds 2. This defines the Mandelbrot set: the set of z-values for which the sequence $|u_n|$ remains bounded below 2.

Table 9.1. The first six members of the sequence defined by the iterative formula Eq. (9.1) with $z = -1.0+0.5i$.

n	u_n	$\|u_n\|$
0	0.0	0.0
1	$-1.0 + 0.5i$	1.11803
2	$-0.25 - 0.5i$	0.55907
3	$-1.1875 + 0.75i$	1.40451
4	$-0.152344 - 1.28125i$	1.29028
5	$-2.61839 + 0.890381i$	2.76564
6	$5.0632 - 4.16273i$	6.55472

[2]Mandelbrot, B.B. (1982), *The Fractal Geometry of Nature* (W.H. Freeman and Co., San Francisco).

[3]Brooks, R. and Matelski, P. (1981), *The Dynamics of 2-Generator Subgroups of PSL(2,C)* (Princeton University Press), pp. 65–71.

Of course, we can only compute a finite number of terms in the sequence u_n by iterating Eq. (9.1). We cannot actually prove that $|u_n|$ remains bounded for any particular value of z. However, numerical experiments show that if $|u_n|$ is less than 2 after, say 100 iterations, it is unlikely to exceed 2 after millions of iterations. As a practical matter, we can define the Mandelbrot set as the set of z-values for which $u_{100} < 2.0$.

With this in mind, define

$$F(z) = \begin{cases} \text{the smallest } n \text{ for which } |u_n| \geq 2, & \text{if } |u_{100}| \geq 2, \\ 0, & \text{if } |u_{100}| < 2. \end{cases} \tag{9.2}$$

Table 9.1 shows that $F(-1+0.5i) = 5$. The Mandelbrot set is defined by $F(z) = 0$.

The Mandelbrot set can be visualized by covering the complex plane with rectangular tiles. Each tile is assigned a color depending on the value of $F(z)$, where $z = x + iy$ is the center of the tile. The Matplotlib function `pcolormesh()` will produce the plot. Here is an example of the main code:

```
# set up grid
xtiles = 1000     # number of tiles in x direction
ytiles = 1000     # number of tiles in y direction
xmin =  -2.0      # minimum x
xmax =  0.8       # maximum x
ymin =  -1.4      # minimum y
ymax =  1.4       # maximum y
xgrid = np.linspace(xmin,xmax,xtiles+1)
ygrid = np.linspace(ymin,ymax,ytiles+1)

# create array for colors
clr = np.zeros((xtiles,ytiles))

# loop over tiles
for i in range(xtiles):
    for j in range(ytiles):
        xmiddle = 0.5*(xgrid[i]+xgrid[i+1])
        ymiddle = 0.5*(ygrid[j]+ygrid[j+1])
        z = xmiddle + ymiddle*1j     # middle of tile
        clr[i,j] = F(z)              # color of tile

# plot complex plane with colors
plt.pcolormesh(xgrid,ygrid,np.transpose(clr),cmap='jet')
```

Comments:

- Remember, in Python `1j` is the imaginary unit. Do not confuse `1j` with the index `j`.
- The array `xgrid` contains the x-coordinates of the left and right edges of each tile. The size of `xarray` is one more than `xtiles`, which is the number of tiles in the x-direction. Likewise, `ygrid` contains the y-coordinates of the bottom and top edges of each tile, and the size of `ygrid` is one more than the number of tiles in the y-direction.
- The array `clr` contains the values of $F(z)$, which are mapped into colors by `pcolormesh`. The `clr` array is transposed because `pcolormesh` interprets the first index as varying along the y-direction, and the second index as varying along the x-direction.
- The nested `for` statements loop over the tiles. The midpoint of each tile is computed and converted into a complex number z. The color value $F(z)$ is computed in a function definition, not shown.
- The color map is specified by `cmap`, which is an optional argument for the function `pcolormesh`.

Exercise 9.4a

Write a Python function definition for $F(z)$, and use this with the main code above to create a picture of the Mandelbrot set. Experiment with different color maps; for example, *hot*, *seismic*, *magma*, *pink*, *gray*, etc. Figure 9.1 shows the Mandelbrot set using the *gray* color map.

The boundary of the Mandelbrot set is a fractal. As you examine the boundary on smaller and smaller scales, complex features continue to emerge. These features include "islands" and "peninsulas" that resemble miniature copies of the set itself.

You can increase the resolution of the Mandelbrot code by modifying `xtiles` and `ytiles`. You can zoom in on interesting regions by changing `xmin`, `xmax`, `ymin` and `ymax`. As you make these changes, increase the maximum number of iterations of Eq. (9.1). This will ensure that the visualization is accurate.

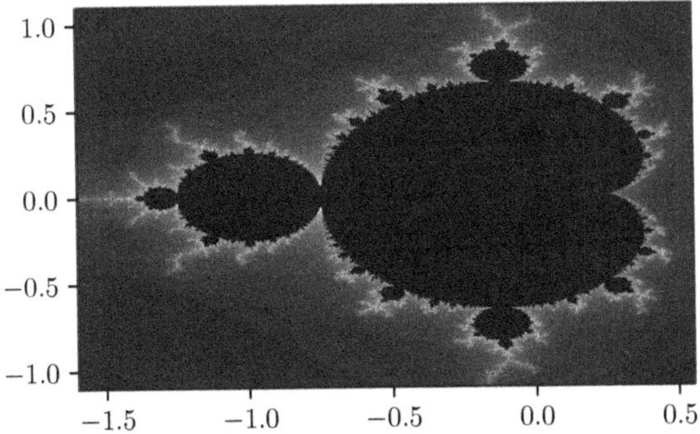

Fig. 9.1. The graph shows the complex z–plane with shading determined by the function $F(z)$ of Eq. (9.2). The dark interior with $F(z) = 0$ is the Mandelbrot set.

Exercise 9.4b

- Create a close-up picture of the region near $z = -0.75 + 0.1i$. This is called "seahorse valley" due to the swirling patterns that resemble seahorses.
- Create a close-up picture of the region near $z = 0.3 + 0.0i$, which is called "elephant valley."
- Find another interesting region of the Mandelbrot set, and create a close-up picture.

The "Burning Ship" fractal is obtained by replacing the iterative formula (9.1) with

$$u_{n+1} = (|\operatorname{Re}(u_n)| + i|\operatorname{Im}(u_n)|)^2 + z^* . \tag{9.3}$$

Here, $\operatorname{Re}(u_n)$ and $\operatorname{Im}(u_n)$ are the real and imaginary parts of u_n, and z^* is the complex conjugate of z. The corresponding NumPy functions are `real()`, `imag()`, and `conjugate()`.

Exercise 9.4c

Modify your Mandelbrot code to produce the Burning Ship fractal. You will want to adjust `xmin`, `xmax`, `ymin`, and `ymax`.

9.5 Colored beads

A child's toy consists of 14 plastic beads that can be snapped together end-to-end to form a circle. The colors of the beads are: 3 red, 3 blue, 2 green, 2 yellow, 2 orange, and 2 purple. If a child connects the beads in random order, what is the probability that no two adjacent beads have the same color?

Exercise 9.5

Write a computer code to simulate the connection of the beads into a circle, in random order. Have your code run at least 100,000 trials to determine the probability that no two adjacent beads have the same color.

Chapter 10

Symbolic Computation with SymPy

Symbolic computation is an important tool for scientists, mathematicians and engineers. Some of the most well-developed systems are Mathematica and Maple. These commercial products are not free. A good alternative is SymPy, a Python library for symbolic computation. This chapter shows, by way of example, some basic capabilities of SymPy.

SymPy is well suited for the Jupyter notebook environment. The examples below are shown as they would appear in a Jupyter notebook.[1]

10.1 Basic SymPy commands

Import SymPy and define variables:

In [1]:

```
import sympy as sp                        # import sympy
x,y,z,b = sp.symbols('x y z beta')        # define variables
x,y,z,b                                   # print variables
```

Out [1]: (x, y, z, β)

[1]In newer versions of Jupyter notebook, the words "In" and "Out" that precede the cell numbers are omitted. For clarity of presentation, "In" and "Out" are used here.

Define functions:

In [2]:
```
f = x**2 - y**2                # define an explicit function
f                              # print f
```
Out [2]: $x^2 - y^2$

In [3]:
```
g = x*(y + z)                  # another explicit function
g                              # print g
```
Out [3]: $x(y+z)$

Factor, expand, and simplify:

In [4]:
```
sp.factor(f)                   # factor f
```
Out [4]: $(x-y)(x+y)$

In [5]:
```
sp.expand(g)                   # expand g
```
Out [5]: $xy + xz$

In [6]:
```
h = f/(x - y) + g              # define h
h                              # print h
```
Out [6]: $x(y+z) + \frac{x^2 - y^2}{x - y}$

In [7]:
```
sp.simplify(h)                 # simplify h
```
Out [7]: $xy + xz + x + y$

Collect, substitute, and coefficients:

In [8]:
```
p = sp.expand((x + z)*(y + z)**2)  # define p
p                                  # print p
```
Out [8]: $xy^2 + 2xyz + xz^2 + y^2z + 2yz^2 + z^3$

In [9]:
```
sp.collect(p,z)            # collect terms in powers of z
```
Out [9]: $xy^2 + z^3 + z^2(x + 2y) + z(2xy + y^2)$

In [10]:
```
p.subs(x,3)                # substitute 3 in place of x
```
Out [10]: $y^2z + 3y^2 + 2yz^2 + 6yz + z^3 + 3z^2$

In [11]:
```
p.subs(z,2*y)          # substitute 2*y in place of z
```
Out [11]: $9xy^2 + 18y^3$

In [12]:
```
p.subs(((x,1),(y,2))) # substitute 1 for x and 2 for y
```
Out [12]: $z^3 + 5z^2 + 8z + 4$

In [13]:
```
p.coeff(z,2)          # coefficient of z^2
```
Out [13]: $x + 2y$

Equations and solutions:

In [14]:
```
f = x**2 - 5*x + 6  # new definition for f
f                   # print f
```
Out [14]: $x^2 - 5x + 6$

In [15]:
```
myeqn = sp.Eq(f,0)  # create the equation f = 0
                    # Eq stands for "Equation" or "Equals"
myeqn               # print myeqn
```
Out [15]: $x^2 - 5x + 6 = 0$

In [16]:
```
solns = sp.solve(myeqn,x)  # solve myeqn for x
solns                      # print solns
```
Out [16]: $[2, 3]$

In [17]:
```
solns[0]     # print first solution (index = 0)
```
Out [17]: 2

In [18]:
```
solns[1]     # print second solution (index = 1)
```
Out [18]: 3

In [19]:
```
myeqn.subs(x,solns[0])    # check first solution
```
Out [19]: True

In [20]:
```
myeqn.subs(x,solns[1])    # check second solution
```
Out [20]: True

Systems of equations:

In [21]:
```
g = 3*x + 2*y           # define function g
eq1 = sp.Eq(g,7)        # define equation g = 7
eq1                     # print eq1
```
Out [21]: $3x + 2y = 7$

In [22]:
```
h = 2*x - y             # Define function h
eq2 = sp.Eq(h,4)        # Define equation h = 4
eq2                     # print eq2
```
Out [22]: $2x - y = 4$

In [23]:
```
solns = sp.solve((eq1,eq2),(x,y))  # solve equations
solns                              # print solns
```
Out [23]: $\{x:\ 15/7,\ y:\ 2/7\}$

In [24]:
```
# The solution is given as a dictionary.
solns[x]      # print dictionary value for x
```
Out [24]: $\frac{15}{7}$

In [25]:
```
solns[y]      # print dictionary value for y
```
Out [25]: $\frac{2}{7}$

In [26]:
```
eq1.subs(((x,solns[x]),(y,solns[y])))    # check eq1
```
Out [26]: True

In [27]:
```
eq2.subs(((x,solns[x]),(y,solns[y])))    # check eq2
```
Out [27]: True

In [28]:
```
# Another example
# Define equations and solve
solns = solve((Eq(x**2 - y**2,0),Eq(x + 2*y,1)),(x,y))
solns         # print solutions
```
Out [28]: $[(-1,\ 1),\ (1/3,\ 1/3)]$

In [29]:
```
# The solution is given as a list of tuples
solns[0]      # print first solution
```
Out [29]: $(-1,\ 1)$

In [30]:
```
solns[1][0]  # print first element of second solution
```
Out [30]: $\frac{1}{3}$

Built-in functions:

In [31]:
```
myroot = sp.sqrt(x*y)       # square roots
myroot                      # print
```
Out [31]: $\sqrt{xy}$

In [32]:
```
mytrig = sp.sin(x)**2 - sp.cos(x)**2  # trig functions
sp.simplify(mytrig)                   # simplify and print
```
Out [32]: $-\cos(2x)$

In [33]:
```
myexp = sp.exp(x)*sp.exp(y)      # exponential functions
sp.simplify(myexp)               # simplify and print
```
Out [33]: e^{x+y}

In [34]:
```
mylog = sp.log(x) + sp.log(y)  # natural logarithms
mylog                          # print
```
Out [34]: $\log(x) + \log(y)$

Assumptions:

In [35]:
```
u = sp.symbols('u')                   # u is complex
v = sp.symbols('v', positive=True)    # v is positive
w = sp.symbols('w', real=True)        # w is real
```

In [36]:
```
sp.sqrt((u*v)**2)              # square root of (u*v)**2
```
Out [36]: $v\sqrt{u^2}$

In [37]:
```
eq1 = sp.Eq(u*(u**2 + 1),0)  # polynomial in u
sp.solve(eq1,u)              # solve eq1
```
Out [37]: $[0, -\mathtt{I}, \mathtt{I}]$

In [38]:
```
eq2 = sp.Eq(w*(w**2 + 1),0)  # same polynomial in w
sp.solve(eq2,w)              # solve eq2
```
Out [38]: $[0]$

Working with numbers:

In [39]:
```
sp.pi                          # pi
```
Out [39]: π

In [40]:
```
sp.pi.evalf()                  # pi to 15 significant figures
```
Out [40]: 3.14159265358979

In [41]:
```
sp.pi.evalf(30)                # pi to 30 significant figures
```
Out [41]: 3.14159265358979323846264338328

In [42]:
```
1/3                               # regular Python calculation
```
Out [42]: 0.3333333333333333

In [43]:
```
sp.Integer(1)/sp.Integer(3)  # symbolic 1/3
```
Out [43]: $\frac{1}{3}$

In [44]:
```
sp.Rational(1,3)                # symbolic 1/3
```
Out [44]: $\frac{1}{3}$

In [45]:
```
# evaluate 1/3 to 12 significant figures
sp.Rational(1,3).evalf(12)
```
Out [45]: 0.333333333333

Exercise 10.1a

Define the polynomial

$$p = (x + y)(x^2 - y)(ax + y)$$

using SymPy.

- Expand p.
- Find the coefficient of x^3.
- Find the coefficient of a.
- Evaluate p at $x = 7$, $y = 3$.

Exercise 10.1b

Use SymPy to compute the following.

- Simplify the function $x - xz/y$.
- Factor the polynomial $x^3 - 2x^2 + x - 2$.
- Evaluate $\sin(e^x)$ at $x = \pi$ to 30 significant figures.

Exercise 10.1c

Solve the system of equations

$$2x^2 - 9y^2 = 1,$$

$$3y - x^2 = 0,$$

using SymPy. Check that the equations are satisfied by each solution.

10.2 Calculus with SymPy

In [1]:

```
import sympy as sp             # import sympy
x,y = sp.symbols('x y')        # define symbols

f = x*sp.cos(y) - y            # define a function f
f                              # print f
```

Out [1]: $x\cos(y) - y$

Derivatives:

In [2]:

```
sp.diff(f,y)        # differentiate f with respect to (wrt) y
```

Out [2]: $-x\sin(y) - 1$

In [3]:

```
sp.diff(f,x,y)  # differentiate f wrt x and y
```

Out [3]: $-\sin(y)$

In [4]:
```
myder = sp.Derivative(f,x) # symbolic derivative df/dx
myder                      # print
```
Out [4]: $\frac{\partial}{\partial x}(x\cos(y) - y)$

In [5]:
```
myder.doit()               # Evaluate symbolic derivative
```
Out [5]: $\cos(y)$

Integrals:

In [6]:
```
sp.integrate(f,x)          # integrate f wrt x
```
Out [6]: $\frac{x^2\cos(y)}{2} - xy$

In [7]:
```
sp.integrate(f,x,y)        # integrate f wrt x and y
```
Out [7]: $\frac{x^2\sin(y)}{2} - \frac{xy^2}{2}$

In [8]:
```
sp.integrate(f,(y,-2,3))   # integrate f wrt y from -2 to 3
```
Out [8]: $x\sin(3) + x\sin(2) - \frac{5}{2}$

In [9]:
```
# integrate f wrt y from -2 to 3, wrt x from -1 to 1
sp.integrate(f,(y,-2,3),(x,-1,1))
```
Out [9]: -5

In [10]:
```
# symbolic integral of f wrt y
myint1 = sp.Integral(f,y)
myint1                     # print
```
Out [10]: $\int (x\cos(y) - y)\, dy$

In [11]:
```
myint1.doit()              # evaluate symbolic integral
```
Out [11]: $x\sin(y) - \frac{y^2}{2}$

In [12]:
```
# symbolic integral of f wrt y from -1 to 1
myint2 = sp.Integral(f,(y,-1,1))
myint2                    # print
```

Out [12]: $\int_{-1}^{1} (x\cos(y) - y)\, dy$

In [13]:
```
myint2.doit()             # evaluate symbolic integral
```

Out [13]: $2x\sin(1)$

Differential equations:

In [14]:
```
F = sp.Function('F')        # Define F as a symbolic function
t = sp.symbols('t')         # define t as a variable
```

In [15]:
```
sp.diff(F(t),t)             # derivative of F(t) wrt t
```

Out [15]: $\frac{d}{dt}F(t)$

In [16]:
```
# create a differential equation
myde = sp.Eq(sp.diff(F(t),t,t) - F(t), sp.exp(t))
myde                        # print
```

Out [16]: $-F(t) + \frac{d^2}{dt^2}F(t) = e^t$

In [17]:
```
sp.dsolve(myde,F(t))        # solve the differential equation
```

Out [17]: $F(t) = C_2e^{-t} + \left(C_1 + \frac{t}{2}\right)e^t$

Limits and series:

In [18]:
```
f = sp.sin(2*x)/x           # define f
f                           # print f
```

Out [18]: $\frac{\sin(2x)}{x}$

In [19]:
```
sp.limit(f,x,0)             # limit of f(x) as x goes to 0
```

Out [19]: 2

In [20]:
```
# Taylor series for f(x) about x=0 through order x**7
myseries = sp.series(f,x,0,8)
myseries                          # print
```

Out [20]: $2 - \frac{4x^2}{3} + \frac{4x^4}{15} - \frac{8x^6}{315} + \mathcal{O}(x^8)$

In [21]:
```
myseries.removeO()      # remove the O(x**8) symbol
```

Out [21]: $\frac{8x^6}{315} + \frac{4x^4}{15} - \frac{4x^2}{3} + 2$

Exercise 10.2a

Consider the function

$$f(x, y) = \sin(x + y)\cos(2x - y).$$

Using SymPy,

- Differentiate $f(x, y)$ with respect to x and simplify the result.
- Integrate $f(x, y)$ with respect to x and simplify the result.
- Integrate $f(x, y)$ with respect to x from $x = -1$ to $x = 1$, substitute the value $y = 2$, and evaluate the answer numerically to 8 significant figures.

Exercise 10.2b

Solve the differential equation

$$x^2 \frac{d^2}{dx^2}F(x) - x\frac{d}{dx}F(x) - 3F(x) = 4x^3$$

using SymPy.

Exercise 10.2c

With SymPy, compute the Taylor series for

$$f(x) = \ln(x + \cos(x)),$$

about $x = 0$. Include terms up to x^7. Remove the $\mathcal{O}(x^8)$ symbol and evaluate the expression at $x = 1/2$. Compare the series approximation to $f(1/2)$.

10.3 Plotting graphs with SymPy

In [1]:
```
import sympy as sp                # import sympy
x,y,u = sp.symbols('x y u')       # define symbols
```

In [2]:
```
# plot 2*sin(x) and sqrt(x/3) for x from 0 to 10
sp.plot(2*sin(x), sqrt(x/3), (x,0,10))
```

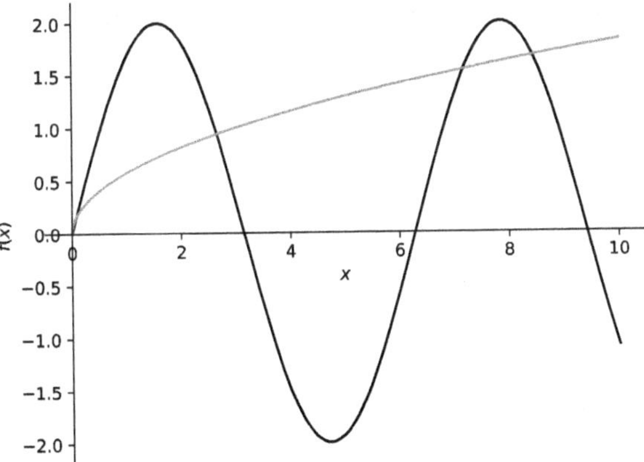

In [3]:
```
# parametric plot of x = sqrt(u), y = cos(u)
# for u from 0 to 10
sp.plot_parametric((sqrt(u),cos(u)), (u,0,10))
```

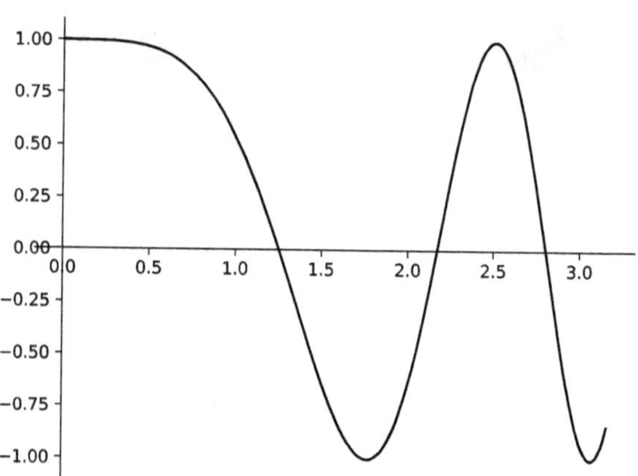

Exercise 10.3a

Use SymPy to plot (on the same graph) the functions $\sin(x)$ and $\sin((\pi/2)\sin(x))$ over the domain $-2\pi \leq x \leq 2\pi$.

Exercise 10.3b

A cycloid is the path made by a point on a circle as the circle rolls along the x-axis. For a circle of radius R, a cycloid is defined parametrically by the equations

$$x = R(t - \sin t),$$
$$y = R(1 - \cos t).$$

Use SymPy to plot the cycloid with $R = 1$.

10.4 Linear algebra with SymPy

In [1]:
```
import sympy as sp                  # import sympy
x,y = sp.symbols('x y')             # define symbols
```

Matrices:

In [2]:
```
A = sp.Matrix([[1,3],[2,5]])  # Define a 2x2 matrix A
A                             # print A
```

Out [2]:
$$\begin{bmatrix} 1 & 3 \\ 2 & 5 \end{bmatrix}$$

In [3]:
```
A[1,0]          # second row, first column
```
Out [3]: 2

In [4]:
```
A[0,1]          # first row, second column
```
Out [4]: 3

In [5]:
```
det(A)          # determinant of A
```
Out [5]: -1

In [6]:
```
A.inv()         # inverse of A
```
Out [6]: $\begin{bmatrix} -5 & 3 \\ 2 & -1 \end{bmatrix}$

In [7]:
```
sp.eye(3)       # 3x3 identity matrix
```
Out [7]: $\begin{bmatrix} 1 & 0 & 0 \\ 0 & 1 & 0 \\ 0 & 0 & 1 \end{bmatrix}$

In [8]:
```
B = sp.Matrix([[3*x,1],[2,-y]])  # another matrix
B                                # print
```
Out [8]: $\begin{bmatrix} 3x & 1 \\ 2 & -y \end{bmatrix}$

In [9]:
```
A*B                     # multiply matrices
```
Out [9]: $\begin{bmatrix} 3x+6 & 1-3y \\ 6x+10 & 2-5y \end{bmatrix}$

In [10]:
```
A - B                   # subtract (or add) matrices
```
Out [10]: $\begin{bmatrix} 1-3x & 2 \\ 0 & y+5 \end{bmatrix}$

Matrix equations:

In [11]:
```
b =  sp.Matrix([[2],[3]])    # column vector
b                            # print b
```
Out [11]: $\begin{bmatrix} 2 \\ 3 \end{bmatrix}$

In [12]:
```
v = A.solve(b) # solve the matrix equation A*v = b for v
v              # print v
```

Out [12]: $\begin{bmatrix} -1 \\ 1 \end{bmatrix}$

Vectors:

In [13]:
```
v1 = sp.Matrix([1,2,3])   # define a vector
v2 = sp.Matrix([4,5,6])   # another vector
v2                        # print v2
```

Out [13]: $\begin{bmatrix} 4 \\ 5 \\ 6 \end{bmatrix}$

In [14]:
```
v1.dot(v2)        # dot product v1.v2
```

Out [14]: 32

In [15]:
```
v1.cross(v2)      # cross product v1 x v2
```

Out [15]: $\begin{bmatrix} -3 \\ 6 \\ -3 \end{bmatrix}$

Exercise 10.4a

Use SymPy to define the matrices

$$M = \begin{pmatrix} 3 & -5 & 2 \\ -1 & 4 & 4 \\ 2 & 3 & -2 \end{pmatrix},$$

$$N = \begin{pmatrix} 1 & 3 & -2 \\ 2 & 5 & -3 \\ 4 & -3 & -2 \end{pmatrix}.$$

Compute the product MN. Find the determinant of MN and the inverse of MN.

Exercise 10.4b

Let

$$A = \begin{pmatrix} 2 & 2 & -4 \\ -1 & 2 & -3 \\ -2 & 1 & 1 \end{pmatrix},$$

$$b = \begin{pmatrix} 3 \\ 5 \\ 1 \end{pmatrix}.$$

With SymPy, solve the matrix equation $Av = b$ for the vector v.

Exercise 10.4c

Use SymPy to define the vector

$$v_1 = \begin{pmatrix} a_1 \\ b_1 \\ c_1 \end{pmatrix},$$

along with two more vectors v_2 and v_3. Verify the triple product identity $v_1 \times (v_2 \times v_3) = (v_1 \cdot v_3)v_2 - (v_1 \cdot v_2)v_3$.

10.5 From SymPy to NumPy

In [1]:

```
import sympy as sp                  # import sympy
import numpy as np                  # import numpy
x = sp.symbols('x')                 # define variables
```

In [2]:

```
f = sp.sin(sp.log(x))               # define a sympy function
f                                   # print
```

Out [2]: $\sin(\log(x))$

In [3]:

```
f = sp.lambdify(x,f,"numpy")        # convert to numpy function
f(2)                                # evaluate f at 2
```

Out [3]: 0.6389612763136348

In [4]:
```
x = np.linspace(2,5,4)          # create array of x values
f(x)                             # evaluate f at x
```

Out [4]: array([0.63896128, 0.89057704, 0.98302774, 0.99925351])

Exercise 10.5

Use SymPy to compute

$$f(x) = \int x \sin^2(x)\, dx.$$

Convert $f(x)$ to a NumPy function, then use Matplotlib to plot the function.

10.6 Printing with SymPy

A final word on printing with SymPy. When we type a variable name such as `f` (or a command such as `factor(f)`) into the last line of a Jupyter notebook cell, Python responds by printing the result to the screen. You can always use the `print()` function to tell Python to print. However, the output might be less readable when you use `print()` explicitly. For example, consider the function $f = x^2 - y^2$. The output from `f` is

$$x^2 - y^2,$$

whereas the output from `print(f)` is

$$\texttt{x**2 - y**2}.$$

SymPy also has a "pretty print" feature. The command `pprint(f)` produces

$$\mathtt{x}^2 - \mathtt{y}^2.$$

These results may differ on different platforms.

Chapter 11

Root Finding

11.1 Motion with air resistance

Consider a ball of mass m, tossed straight up into the air. Let y denote the height of the ball above ground level. The ball's velocity is $\dot{y}$, where the dot denotes a derivative with respect to time t. The force of gravity acting on the ball is $F_{\text{grav}} = -mg$, where g is the acceleration due to gravity. Air resistance also acts on the ball, exerting a force in the direction opposite to the velocity. For low velocities, this force can be approximated as $F_{\text{air}} = -mk\dot{y}$, where k is a positive constant. (The factor of m is included for later convenience.)

Newton's second law for the ball is $F_{\text{grav}} + F_{\text{air}} = m\ddot{y}$, where $\ddot{y}$ is the ball's acceleration. This yields a simple second-order differential equation

$$\ddot{y} + k\dot{y} + g = 0, \tag{11.1}$$

for the height y as a function of time t. Standard techniques can be used to derive the solution,

$$y(t) = \left(v_0/k + g/k^2\right)\left(1 - e^{-kt}\right) - (g/k)t, \tag{11.2}$$

where v_0 is the initial velocity. The ball is launched from ground level, so the initial height is $y(0) = 0$.

Exercise 11.1a

Use SymPy to check the result (11.2) by substituting $y(t)$ into the differential equation (11.1). Also verify that the initial velocity is $\dot{y}(0) = v_0$.

How long does the ball remain in the air? In other words, for what value of time t (in addition to $t = 0$) is the height equal to zero? The answer is found by setting $y(t) = 0$ in Eq. (11.2) and solving for t. However, it's not possible to solve for t analytically in terms of standard functions like powers, trig functions, logarithms, etc. We must resort to numerical methods to find the time in air.

One way to solve $y(t) = 0$ numerically is to start with an initial guess, then increase (or decrease) the guess in small steps until the function $y(t)$ changes sign. Since $y(t)$ has opposite signs for the last two guesses, the function $y(t)$ must vanish for some value of t between those guesses. We can approximate the time where $y(t)$ vanishes by taking the average of the last two guesses.

Exercise 11.1b

Let $g = 9.8$, $k = 0.2$, and $v_0 = 35.0$ (in SI units) in Eq. (11.2). Write a Python program that uses the algorithm described above to solve $y(t) = 0$. Start with an initial guess close to 0 and use a time step size of 0.00001. Use your result to determine how fast the ball is moving when it reaches the ground.

11.2 The general problem

We are often faced with the problem of finding the solution (or solutions) of an algebraic equation of the form $f(x) = 0$. The air resistance problem has this form, but with a simple change of variable names (y instead of f, t instead of x.)

The solutions of $f(x) = 0$ are called the "roots" of the function $f(x)$. In this chapter we develop numerical techniques for finding roots. The goal is to find a root to within some tolerance ϵ. That is, the root should be accurate to within $\pm\epsilon$.

The straightforward method for root finding, described in the previous section, is usually inefficient. To obtain an accurate answer you must use a very small step size, and this requires very many steps (unless your initial guess is really good). There are better methods.

11.3 The bisection method

Bisection is a simple method of finding the roots of a continuous function $f(x)$. Start by assuming a root exists between $x = x_L$ and $x = x_R$. If $f(x_L)$ and $f(x_R)$ have opposite signs, then there must be at least one root between x_L and x_R. Next, compute the midpoint of the interval $x_M = (x_L + x_R)/2$. Use this value to replace either x_L or x_R, depending on the relative signs of $f(x_L)$, $f(x_M)$, and $f(x_R)$. Repeat this process until x_L and x_R are close enough that the root (which must be between x_L and x_R) can be reasonably approximated by the midpoint $x_M = (x_L+x_R)/2$. The idea is depicted in Fig. 11.1.

Here is the step-by-step procedure:

(1) Choose values for x_L and x_R, with $x_L < x_R$.
(2) Verify that $f(x_L)$ and $f(x_R)$ have opposite signs. If not, adjust your choices for x_L and x_R.
(3) Compute the midpoint $x_M = (x_L + x_R)/2$.
(4) If $f(x_M)$ and $f(x_L)$ have the *same* sign, then x_M must be smaller than the root we're looking for. Replace the value of x_L with x_M.

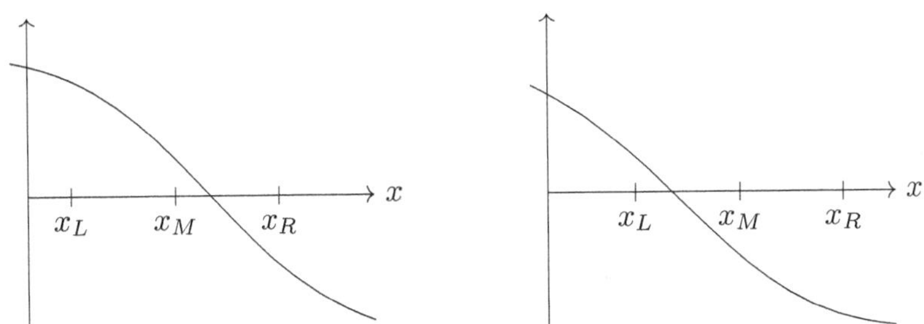

Fig. 11.1. The bisection method. The actual root is where the curve crosses the x-axis. Left panel: $f(x_M)$ has the same sign as $f(x_L)$, so x_M is smaller than the actual root. For the next iteration, x_M becomes the new x_L. Right panel: $f(x_M)$ has the same sign as $f(x_R)$, so x_M is larger than the actual root. For the next iteration, x_M becomes the new x_R.

(5) If $f(x_M)$ and $f(x_R)$ have the same sign, then x_M must be *larger* than the root we're looking for. Replace the value of x_R with x_M.
(6) Repeat steps 3, 4, and 5 until $x_R - x_L < 2\epsilon$, where ϵ is some chosen tolerance.
(7) Take the midpoint $(x_L + x_R)/2$ as your final answer. Since the actual root is between x_L and x_R, and $x_R - x_L < 2\epsilon$, the midpoint must be within $\pm\epsilon$ of the actual root.

The bisection method is robust, meaning it almost always succeeds in finding a root.

Exercise 11.3a

Use bisection to solve the equation

$$\exp(-x^2) - x = 0,$$

with a tolerance of $\epsilon = 0.0001$.

Exercise 11.3b

The bisection method can converge on a vertical asymptote, as well as a root. Consider the function $\tan(x) - x$. Plot this function and apply the bisection method with (initially) $x_L = 1$ and $x_R = 2$. Does this give a root or a vertical asymptote? Use bisection to find the other roots and vertical asymptotes in the domain $0 < x < 6$.

11.4 Newton's method

Newton's method (also known as the Newton–Raphson method) is much more efficient than the bisection method—typically it reaches the answer with a comparable tolerance in fewer steps. However, Newton's method is less robust than bisection, and it can be more difficult to apply because it requires knowledge of the derivative of $f(x)$.

Let x_{old} denote an initial guess for a root of $f(x)$. Consider the Taylor series expansion of $f(x)$ about x_{old}:

$$f(x) = f(x_{\text{old}}) + f'(x_{\text{old}})(x - x_{\text{old}}) + \cdots . \tag{11.3}$$

Here, $f'(x_{\text{old}})$ is the derivative of $f(x)$ evaluated at x_{old}. The unwritten terms in Eq. (11.3) are proportional to $(x - x_{\text{old}})^2$, $(x - x_{\text{old}})^3$, etc. Let's assume for the moment that x is an exact root of the function, so that $f(x) = 0$. If the initial guess x_{old} is close to the root x, then the unwritten terms in Eq. (11.3) will be small. Dropping these terms and setting $f(x) = 0$, Eq. (11.3) becomes

$$0 = f(x_{\text{old}}) + f'(x_{\text{old}})(x - x_{\text{old}}). \tag{11.4}$$

Solving for x, we find

$$x = x_{\text{old}} - \frac{f(x_{\text{old}})}{f'(x_{\text{old}})}. \tag{11.5}$$

This equation gives us the result x for the root. This result is not exact, because we approximated Eq. (11.3) by dropping the higher order terms.

Rather than calling the result x, let's call it x_{new} and write Eq. (11.5) as

$$x_{\text{new}} = x_{\text{old}} - \frac{f(x_{\text{old}})}{f'(x_{\text{old}})}. \tag{11.6}$$

As noted above, x_{new} is only an approximation to the actual root. Nevertheless, x_{new} is usually a better approximation than our initial guess x_{old}.

Now iterate the process. Let the new approximation x_{new} become the old approximation x_{old} and apply the formula (11.6) again. With each iteration, the new x value is typically closer to the actual root than the old x value. To obtain an answer that is accurate to within some tolerance $\pm\epsilon$, keep iterating until $f(x_n + \epsilon)$ and $f(x_n - \epsilon)$ have opposite signs. This ensures that an actual root is between $x_n + \epsilon$ and $x_n - \epsilon$.

Newton's method is depicted in Fig. 11.2. The slope $f'(x_{\text{old}})$ is the "rise" $f(x_{\text{old}})$ over the "run" $x_{\text{old}} - x_{\text{new}}$; that is,

$$f'(x_{\text{old}}) = \frac{f(x_{\text{old}})}{x_{\text{old}} - x_{\text{new}}}. \tag{11.7}$$

This can be rearranged to give Eq. (11.6).

Newton's method will fail to find a root if any of the x_{old} values coincide with an extremum of $f(x)$. At an extremum, the derivative $f'(x_{\text{old}})$ vanishes and x_{new} becomes infinite.

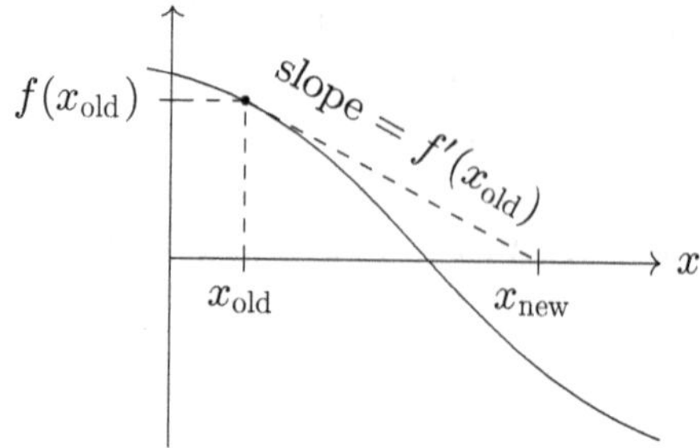

Fig. 11.2. For Newton's method, the new approximation x_{new} is obtained from x_{old} by assuming the function is a straight line with slope $f'(x_{old})$.

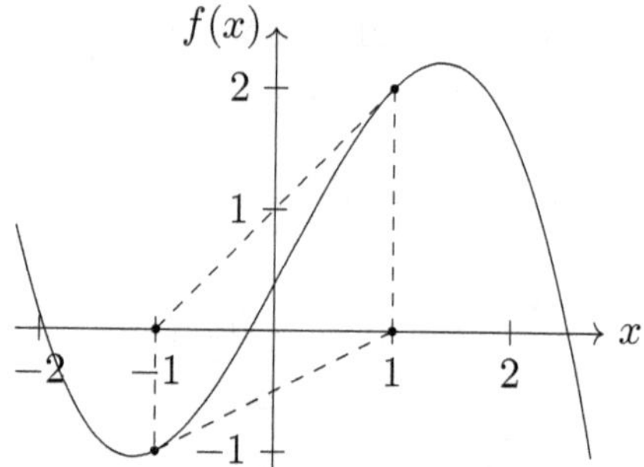

Fig. 11.3. Newton's method applied to the function $f(x) = (-3x^3 + x^2 + 15x + 3)/8$. The initial guess is $x_{old} = 1$. Since $f(1) = 2$ and $f'(1) = 1$, this yields $x_{new} = -1$. Iterating with $x_{old} = -1$, we have $f(-1) = -1$ and $f'(-1) = 1/2$, which gives $x_{new} = 1$.

Newton's method can fail in other ways. Consider, for example, the function $f(x) = (-3x^3 + x^2 + 15x + 3)/8$. If we choose $x_{old} = 1$ as our initial guess, the sequence of old and new estimates of the root will simply switch between 1 and −1. See Fig. 11.3.

Exercise 11.4a

Use Newton's method to find the root of $f(x) = x\ln(1 + x^2) + \cos x$ with a tolerance of $\epsilon = 10^{-5}$.

Exercise 11.4b

The equation

$$x^5 + tx^3 + x - t^3 = 0$$

defines x as a function of t; that is, $x(t)$. For a given value of t, we can use Newton's method to find the corresponding x. Using this strategy, write a code that will graph the function $x(t)$ for $-3.0 \leq t \leq 3.0$.

Exercise 11.4c

In the year AD 60, the Greek mathematician Hero of Alexandria described a method for computing $\sqrt{S}$, where S is a positive number. This is now called Heron's method. Start with an initial guess $x_{\text{old}} > 0$, then iterate the formula

$$x_{\text{new}} = \frac{1}{2}\left(x_{\text{old}} + \frac{S}{x_{\text{old}}}\right).$$

As the number of iterations increases, x_{new} approaches $\sqrt{S}$.

- Show that Heron's method is equivalent to Newton's method for finding the positive root of $f(x) = x^2 - S$.
- Write a code that uses Heron's method to compute square roots with a tolerance of $\pm 10^{-8}$.

11.5 Execution time

You can find out how long your codes take to execute using the `time()` function from the `time` library. Begin by importing the library:

```
import time as tm
```

Add `begintime = tm.time()` at the beginning of your code, and `endtime = tm.time()` at the end. The execution time is `endtime - begintime`.

Exercise 11.5a

Use the `time` library to determine the execution time for your code from Exercise 11.1b.

Exercise 11.5b

Let $g = 9.8$, $k = 0.2$ and $v_0 = 35.0$ (in SI units) in Eq. (11.2). Write a code using the bisection method to find the time in air for the ball. Your answer should have a tolerance of $\epsilon = 0.00001$. (That is, it should be accurate to within ± 0.00001.) Use `time` to find the execution time of your code.

Exercise 11.5c

Repeat the previous exercise, but replace the bisection method with Newton's method.

11.6 Multiple roots

What happens if the function $f(x)$ has more than one root?

The bisection method assumes that the initial values of x_L and x_R are chosen such that $f(x_L)$ and $f(x_R)$ have opposite signs. This implies an odd number of roots between x_L and x_R. If there is just one root in the interval $x_L < x < x_R$, then the bisection method will find that root. If there are multiple roots between x_L and x_R, your bisection code will probably find one of the roots, depending on how it is written.

Exercise 11.6a

Graph the cubic polynomial $f(x) = x^3 - 4x^2 + 2$. Use bisection to find all three roots to within a tolerance of $\epsilon = 0.0001$. What happens if there are two roots between the initial x_L and x_R? Three roots?

For a function with more than one root, Newton's method typically finds the root that is closest to the initial guess.

Exercise 11.6b

Graph the function $f(x) = \cos(2x) - x/2$ and estimate the values of the roots. Use Newton's method to find these roots to a tolerance of $\epsilon = 0.0001$. Show that Newton's method finds the closest root as long as your initial guess is within a few tenths of that root. Which root does Newton's method find when your initial guess is about half way between two roots? If your initial guess is far away from all roots, does Newton's method always find the closest root?

11.7 SciPy optimize

The Python library `scipy.optimize` has built-in functions for root finding. The following code finds the roots of $f(x) = \cos(x) + \sin(x)$:

```
import scipy.optimize as so
import numpy as np

def fxn(x):
    return np.cos(x) + np.sin(x)

# Find root of fxn with initial guess 3 and tolerance 0.0001
rootinfo = so.root(fxn, 3, tol=0.0001)
print(f"root = {rootinfo.x[0]:.4f}")
```

The function `root()` returns a lot of information; here, the information is stored in a variable called `rootinfo`. The numerical approximation to the root is contained in the NumPy array `rootinfo.x`. In this example, `rootinfo.x[0]` is the first (and only) element in the array. Other information returned by `root()` includes `rootinfo.fun`, which is the value of $f(x)$ at the approximate root and `rootinfo.message`, which tells whether or not the algorithm was successful in finding a root.

The output of this code is `root = 2.3562`. There are other roots which can be found using other initial guesses.

Exercise 11.7a

Consider the function

$$f(x) = \frac{1}{5}x^{3/2} + \frac{1}{9}x + 4\cos(x) - 3, \quad x \geq 0.$$

Write a code that will plot a graph of $f(x)$ and use the `scipy.optimize` function `root()` to find the roots. Experiment with different values for the tolerance.

The function `root()` can also be used to solve systems of equations. Consider, for example, the two equations

$$x^2 - y - 2 = 0, \tag{11.8a}$$

$$y^2 - x - 1 = 0, \tag{11.8b}$$

in two unknowns x and y. After importing `scipy.optimize` as `so`, this problem can be solved with the code

```
def fxns(variables):
    x,y = variables
    eq1 = x**2 - y - 2
    eq2 = y**2 - x - 1
    return [eq1,eq2]

# Find root using initial guess x=1, y=2 and tolerance 0.001
rootinfo = so.root(fxns, [1,2], tol=0.001)
print(rootinfo.x)
```

Exercise 11.7b

The system

$$x^3 - y + 1 = 0$$

$$x^2 + xy - y = 0$$

has two real solutions. Use `root()` to find them.

Exercise 11.7c

Use `root()` to solve

$$x - y^4/10 - z^4/10 = 0,$$
$$y - x^4/10 - z^4/10 = 0,$$
$$x^2 + y^2 + z^4/10 - 1 = 0.$$

There are two real solutions.

Chapter 12

Curve Fitting and Interpolation

12.1 Expanding Universe

In 1929, Edwin Hubble analyzed the observational data for a group of 24 nebulae (large clouds of gas and dust) beyond the Milky Way galaxy.[1] For these few nebulae the data were thought to be reliable enough to provide an estimate of the distance to the nebula, d, as well as the radial velocity of the nebula, v. The data are given in Table 12.1, and plotted in Fig. 12.1. Although the data are very scattered, they show a rough trend: galaxies that are farther away from us are receding from us at a faster rate.

Hubble's law is the assertion that, on average, the recession velocity v is proportional to the distance d. That is, the data shown in Fig. 12.1 represent a straight line[2]

$$v = H_0 d. \tag{12.1}$$

The slope H_0 is called the *Hubble constant.* We need a rational way to find the slope from the scattered data.

Currently, the best measurements of the Hubble constant yield values around $H_0 = 70\,(\text{km/s})/\text{Mpc}$. In other words, galaxies that

---

[1]Hubble, E. (1929). A relation between distance and radial velocity among extragalactic nebulae, *Proc. Natl. Acad. Sci.* U.S.A. **15**, 3, 168–173.

[2]The data are affected by the motion of our solar system. Hubble applied a correction to the raw data to account for this motion. The corrected data is still quite scattered, but slightly more suggestive of a straight-line relationship.

Table 12.1. Data from Hubble's 1929 study of extra-galactic nebulae. Distances d are in mega-parsecs (Mpc) and velocities v are in kilometers per second (km/s). One Mpc is equal to 3.26×10^6 light-years.

d (Mpc)	v (km/s)	d (Mpc)	v (km/s)	d (Mpc)	v (km/s)
0.032	170	0.5	270	1.1	450
0.034	290	0.63	200	1.1	500
0.214	−130	0.8	300	1.4	500
0.263	−70	0.9	−30	1.7	960
0.275	−185	0.9	650	2.0	500
0.275	−220	0.9	150	2.0	850
0.45	200	0.9	500	2.0	800
0.5	290	1.0	920	2.0	1090

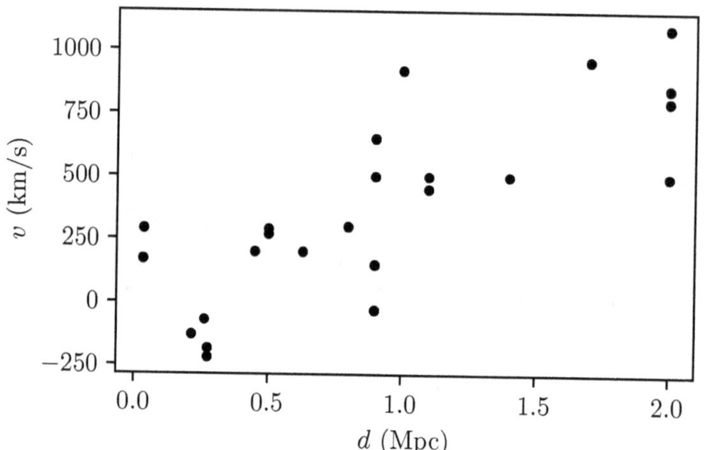

Fig. 12.1. The data from Table 12.1 suggest that the recession velocity between galaxies is greater for galaxies that are farther apart.

are 1 Mpc apart are (on average) receding away from each other at 70 km/s. Galaxies that are 2 Mpc apart are receding away from each other at about 140 km/s.

Exercise 12.1

The data from Table 12.1 is contained in the file *HubbleData.txt*. See the Appendix. Write a code that will read this data and reproduce the graph in Fig. 12.1.

12.2 Least squares method

How do we find the slope H_0 from the scattered data of Fig. 12.1? One way to determine H_0 is the *least squares method.*

Let d_i and v_i denote the data values from Table 12.1, where the index i ranges from 0 through 23. Thus, $d_0 = 0.032$, $v_0 = 170$, etc. With the least squares method, we look for the value of H_0 that minimizes the sum of squares of differences between $H_0 d_i$ and v_i. This is depicted in Fig. 12.2. To be precise, we define the sum

$$S = \sum_{i=0}^{23} (H_0 d_i - v_i)^2, \tag{12.2}$$

and minimize S with respect to H_0.

Let's adopt a simpler notation by dropping the limits on the sum and writing $S = \sum (H_0 d_i - v_i)^2$. Note that S is a quadratic function of H_0. The minimum of S is found by setting the derivative to zero:

$$\frac{dS}{dH_0} = \sum 2H_0(d_i)^2 - \sum 2v_i d_i = 0. \tag{12.3}$$

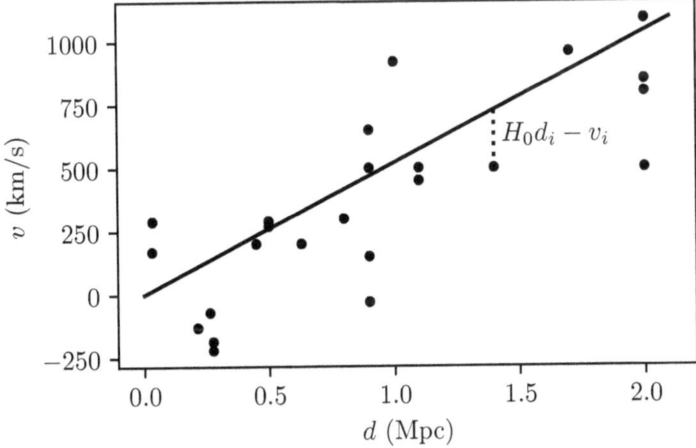

Fig. 12.2. $H_0 d_i - v_i$ is the "distance" between the ith data point and the line $v = H_0 d$. The least squares method determines the value of H_0 that minimizes the sum of the squares of these distances.

Since H_0 is a constant, independent of i, we can bring this factor outside the summation sign and solve; the result is

$$H_0 = \frac{\sum v_i d_i}{\sum (d_i)^2} . \tag{12.4}$$

With this value of H_0, the line $v = H_0 d$ is a "best fit" for the data in Table 12.1.

Exercise 12.2

Extend your code from Exercise 12.1 to determine the Hubble constant H_0 using the least squares method. Plot the best-fit line $v = H_0 d$ and the data points on the same graph.

Hubble's result for H_0 is rather far from the modern-day value of around 70 (km/s)/Mpc. In 1929, the observational errors were quite large. Nevertheless, Hubble's analysis is historically important because it represents the first observational evidence that the Universe is expanding.

12.3 Best-fit line

Let x_i, y_i denote a set of data with $i = 0, \ldots, N-1$. We can apply the least squares method to find a function $f(x)$ that fits this data. The first step is to choose a form for $f(x)$. There are many situations in which the expected relationship between the variables x and y is linear. So let's assume a linear form for the unknown function, $f(x) = ax + b$. Hubble's law (12.1) is the special case in which the y-intercept b vanishes.

The sum of squares of differences between the function values $f(x_i) = ax_i + b$ and the data values y_i is

$$S = \sum_{i=0}^{N-1} (ax_i + b - y_i)^2 . \tag{12.5}$$

S is a quadratic function of the coefficients a and b, and its minimum satisfies $\partial S/\partial a = 0$ and $\partial S/\partial b = 0$. Explicitly, we have

$$\frac{\partial S}{\partial a} = 2\sum (ax_i + b - y_i)x_i = 0, \tag{12.6a}$$

$$\frac{\partial S}{\partial b} = 2\sum (ax_i + b - y_i) = 0. \tag{12.6b}$$

Limits on the summation signs have been dropped for notational simplicity. Since a and b are constants, independent of the summation index i, these results simplify to

$$a\sum x_i^2 + b\sum x_i = \sum x_i y_i \tag{12.7a}$$

$$a\sum x_i + bN = \sum y_i, \tag{12.7b}$$

where $\sum b = Nb$. This is a system of linear equations for the two unknowns, the slope a and the y-intercept b of the line that best fits the data.

Exercise 12.3a

Solve the two equations (12.7) for a and b in terms of $\sum x_i^2$, $\sum x_i$, $\sum x_i y_i$, and $\sum y_i$. (Hint: Invent a simplified notation for the sums, such as Sxx for $\sum x_i^2$ and Sxy for $\sum x_i y_i$. Then use SymPy to solve for a and b.)

Exercise 12.3b

Import the data file *LineData.txt* (see the Appendix). Apply the least squares method to fit the data to a straight line, using the formulas for the slope a and y-intercept b from the previous exercise. Make a plot that shows the data points and the best-fit line.

12.4 Best-fit curve

Least squares analysis can be applied to any class of functions $f(x)$. For example, consider the data from the graph of Fig. 12.3. The data

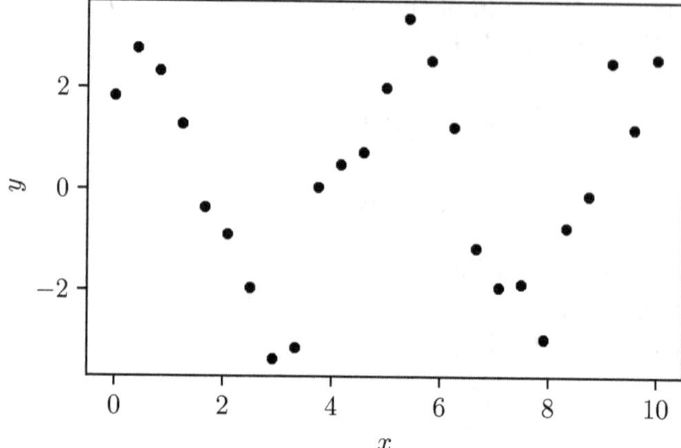

Fig. 12.3. The data suggest a sin–curve relationship, but the amplitude, frequency and phase are unknown.

appear to follow a sin–curve relationship, but we don't know the amplitude, frequency or phase. We can carry out a least squares analysis to fit this data to a function of the form

$$f(x) = a\sin(bx + c). \tag{12.8}$$

The goal is to find the parameter values a, b, and c that minimize

$$S = \sum_{i=1}^{N} (a\sin(bx_i + c) - y_i)^2, \tag{12.9}$$

by solving the equations $\partial S/\partial a = 0$, $\partial S/\partial b = 0$ and $\partial S/\partial c = 0$. These equations are nonlinear and cannot be solved analytically; they must be solved numerically. Fortunately, the function `curve_fit()` in the `scipy.optimize` library will do the work for us. Begin by importing `scipy.optimize` as `so` and creating the function definition

```
def func(x,a,b,c):
    return a*np.sin(b*x + c)
```

The first argument is the independent variable (in this case x), and the remaining arguments are the parameters (in this case a, b, and c). Let's assume the data are contained in arrays called `xdata` and `ydata`. Then the command

```
params, cov = so.curve_fit(func,xdata,ydata)
```

will return the least squares values of a, b, and c in the parameter array `params`. The `curve_fit()` function also returns an array `cov` containing the *covariance matrix*. The square root of each diagonal element of `cov` provides an estimate of the standard error in the corresponding parameter value. For example, the parameter a is contained in `params[0]` and its standard error is `sqrt(cov[0,0])`. The standard errors tell us how well the parameter values are constrained by the data, and give us a measure of how well the function fits the data.

Exercise 12.4a

Write a code that will import the data file *SineData.txt* (see the Appendix) and carry out a least squares fit to a function of the form (12.8). Plot a graph showing the data points, as well as the best-fit curve. What values did you get for the amplitude, frequency and phase?

Exercise 12.4b

Write a code to import the data file *PolyData.txt* (see the Appendix). Carry out a least squares fit to each of the three polynomials

$$f(x) = cx + d$$
$$f(x) = bx^2 + cx + d$$
$$f(x) = ax^3 + bx^2 + cx + d$$

Have your code plot the data and each of the three best-fit curves on the same graph.

Exercise 12.4c

Use `curve_fit()` to find the best-fit line for the data from *LineData.txt*. Do your results match those from Exercise 12.3b? What is the standard error in each of the parameter values?

12.5 Data interpolation

Consider the data displayed in Table 12.2 and plotted in Fig. 12.4. What is the value of y at $x = 3.6$? To answer this question, we need to *interpolate* the data.

You might wonder how the data were obtained. Can't we simply measure or observe or compute the value of y at $x = 3.6$? It might not be practical to do so. For example, the data might come from observations of a comet passing near the sun and can't be repeated for years. Or the data might come from a complicated numerical calculation. Each data point in Table 12.2 could take weeks or months to compute. We want to know $y(3.6)$ now.

A familiar example of interpolation is seen on a daily weather map showing isotherms, curves of constant temperature. (Or isobars, curves of constant pressure.) The data for these maps are collected at discrete times from weather stations at discrete locations, spread around the world. The data must be interpolated, first in time then in space, to obtain the isotherms.

Table 12.2. Discrete data obtained from a set of measurements, observations, or simulations.

x	0.4	1.3	3.1	4.1	4.9	5.8	6.9
y	1.7	2.9	2.8	2.0	1.1	0.7	1.3

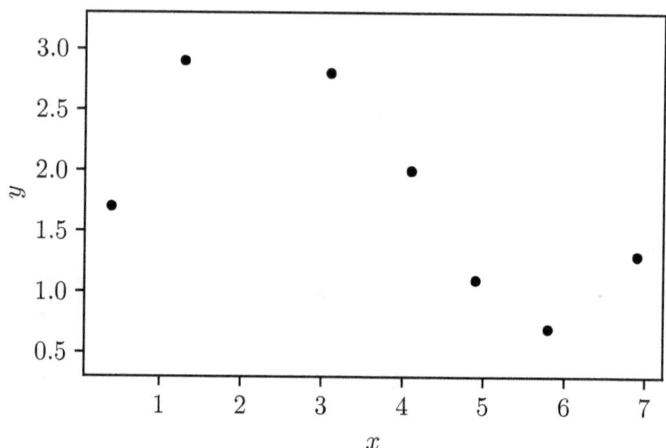

Fig. 12.4. Graph of the data listed in Table 12.2.

Interpolation is appropriate when the known data are relatively accurate, but further data values are difficult to obtain. In the context of interpolation, the data points are often called "knots."

Return to the data of Table 12.2. We can use *linear interpolation* to find an approximate value for $y(3.6)$. With linear interpolation we simply connect the knots with straight lines. More precisely, for every adjacent pair of knots, we find a linear function that passes through the knots. For adjacent knots (x_0, y_0) and (x_1, y_1), the interpolating function is

$$y(x) = \frac{(y_1 - y_0)}{(x_1 - x_0)}(x - x_0) + y_0. \tag{12.10}$$

For the data from Table 12.2, the interpolating function between knots $(3.1, 2.8)$ and $(4.1, 2.0)$ is

$$y(x) = \frac{(2.0 - 2.8)}{(4.1 - 3.1)}(x - 3.1) + 2.8 = 5.28 - 0.8x. \tag{12.11}$$

We can now answer our question: $y(3.6) = 5.28 - (0.8)(3.6) = 2.4$.

Exercise 12.5

The data from Table 12.2 are contained in the file *InterpData.txt*. Write a code to define the function $y(x)$ that linearly interpolates this data. Plot the function along with the data points. Use your function to approximate $y(2.0)$ and $y(6.5)$. (Note: The function $y(x)$ is piecewise continuous, defined by Eq. (12.10) in the intervals between each adjacent pair of knots.)

12.6 Cubic splines

Linear interpolation is not always desirable because the resulting function is not smooth. In particular, it is generally not differentiable at the location of each knot. With linear interpolation we cannot answer questions such as: What is the (approximate) value of the derivative dy/dx at $x = 5.8$?

A better option is to span the gaps between knots using higher order polynomials. We could use quadratic polynomials, but it turns out that cubic polynomials are usually preferred. There is enough freedom in a cubic polynomial to insure that first and second derivatives are continuous at the knots. The interpolating polynomials are called *cubic splines.*

Cubic splines are constructed as follows. For simplicity, let's consider just 3 data points (x_0, y_0), (x_1, y_1), and (x_2, y_2). There are two gaps between these three data points, as shown in Fig. 12.5. We seek a cubic polynomial

$$p_1(x) = a_1 + b_1 x + c_1 x^2 + d_1 x^3 \tag{12.12}$$

to span the first gap, and a cubic polynomial

$$p_2(x) = a_2 + b_2 x + c_2 x^2 + d_2 x^3 \tag{12.13}$$

to span the second gap. We want these polynomials to meet smoothly at the middle knot (x_1, y_1).

To be precise, we need eight conditions on the polynomials (12.12) and (12.13) to determine the eight unknown coefficients a_1, b_1, c_1, d_1, a_2, b_2, c_2, and d_2. To begin, we demand that the polynomials match the data at their endpoints. This provides four equations: $p_1(x_0) = y_0$, $p_1(x_1) = y_1$, $p_2(x_1) = y_1$, and $p_2(x_2) = y_2$. Explicitly, these conditions yield

$$a_1 + b_1 x_0 + c_1 x_0^2 + d_1 x_0^3 = y_0, \tag{12.14a}$$

$$a_1 + b_1 x_1 + c_1 x_1^2 + d_1 x_1^3 = y_1, \tag{12.14b}$$

$$a_2 + b_2 x_1 + c_2 x_1^2 + d_2 x_1^3 = y_1, \tag{12.14c}$$

$$a_2 + b_2 x_2 + c_2 x_2^2 + d_2 x_2^3 = y_2. \tag{12.14d}$$

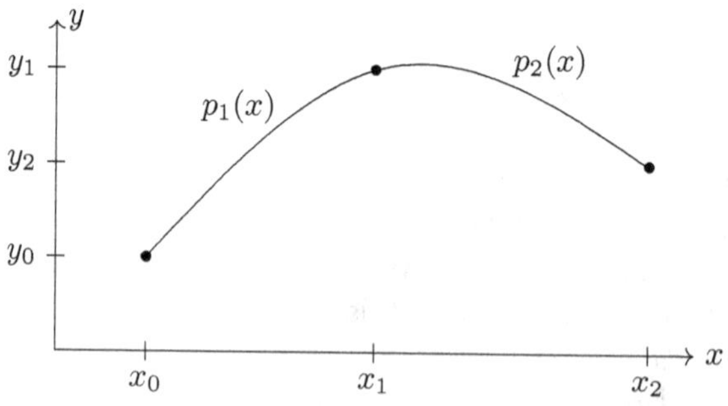

Fig. 12.5. Two cubic spline polynomials span the gaps between three knots.

We also demand that the first and second derivatives of the two polynomials should match at the middle knot, (x_1, y_1). This gives $p_1'(x_1) = p_2'(x_1)$ and $p_1''(x_1) = p_2''(x_1)$, which are explicitly

$$b_1 + 2c_1x_1 + 3d_1x_1^2 = b_2 + 2c_2x_1 + 3d_2x_1^2, \tag{12.14e}$$

$$2c_1 + 6d_1x_1 = 2c_2 + 6d_2x_1. \tag{12.14f}$$

We need two more equations to determine the eight coefficients. The usual choice is to set the second derivatives at the endpoints to zero. That is, $p_1''(x_0) = 0$ and $p_2''(x_2) = 0$, which give

$$2c_1 + 6d_1x_0 = 0, \tag{12.14g}$$

$$2c_2 + 6d_2x_2 = 0. \tag{12.14h}$$

The eight equations (12.14) can be solved for the eight coefficients.

Exercise 12.6

Solve the system of equations (12.14) with the values $(x_0, y_0) = (1.3, 2.9)$, $(x_1, y_1) = (3.1, 2.8)$ and $(x_2, y_2) = (4.1, 2.0)$ from Table 12.2. (Hint: use `SymPy`.) Construct the polynomials $p_1(x)$ and $p_2(x)$ and graph the results. What is $y(3.6)$? What is the derivative dy/dx at $x = 3.1$?

12.7 More data points

The pattern for cubic splines will work with any number of knots. Let's check.

If there are $N + 1$ knots, then there are N gaps between knots. With a cubic polynomial in each gap, there are $4N$ coefficients. We need $4N$ equations to determine these coefficients. Now, each spline should match the data at their endpoints. This provides $2N$ equations. Note that there are $N - 1$ *interior* knots where two splines meet. Setting the first derivative of adjacent splines equal to each other gives $N - 1$ equations. Setting the second derivatives of adjacent splines equal to one another gives another $N - 1$ equations. Finally, we set the second derivative of the first spline equal to zero at its left and, and the second derivative of the last spline equal to zero at its right end. This is a total of $4N$ equations, as desired.

With a bit of effort we could write a computer program to construct cubic splines for an arbitrary number of knots. Fortunately, this has been done for us, and implemented in the library `scipy.interpolate`. Given data arrays `x` and `y`, the command

```
fun = si.CubicSpline(x,y,bc_type='natural',extrapolate=False)
```

(with `scipy.interpolate` imported as `si`) produces a function that interpolates the data with cubic splines. This function, which we have named `fun`, is the *cubic spline interpolator* for the data set. You can evaluate `fun` like any other function. If `z` is a number, `fun(z)` gives the value of `fun` at `z`. If `z` is a NumPy array, `fun(z)` is the NumPy array obtained by evaluating `fun` at each element of `z`.

The option `bc_type='natural'` implements the "natural" boundary conditions discussed previously, namely, the vanishing of second derivatives at the first and last knots. The option `extrapolate=False` tells Python not to extrapolate the data beyond the first and last knots. If you try to evaluate `fun` at a value outside the domain, Python will return `nan` (which stands for "not a number").

Exercise 12.7a

Use `CubicSpline()` to compute the cubic spline approximation to the data given in Table 12.2 and *InterpData.txt*. Plot the cubic spline interpolator along with the data.

Because the cubic spline interpolator (which we call `fun`) is a piecewise polynomial, it is straightforward to differentiate and integrate. The first derivative is given by `fun.derivative()`. The `n`th derivative is obtained by replacing `()` with `(n)`. The indefinite integral of the interpolator is `fun.antiderivative()`. The constant of integration is chosen such that the integral equals zero at the left-end knot.

Exercise 12.7b

Create a data set by sampling the function

$$f(x) = \cos(x) + \sin(2x)$$

for x between $-\pi$ and π. That is, define the arrays

```
x = np.linspace(-np.pi,np.pi,N)
y = np.cos(x) + np.sin(2*x)
```

for some chosen value of `N`. Use `CubicSpline()` to obtain the cubic spline interpolator and compare to the original function $f(x)$ by plotting the difference. Find the derivative of the interpolator and compare to the derivative of $f(x)$ by plotting the difference. Find the integral of the interpolator and compare to the integral of $f(x)$ by plotting the difference. (If needed, adjust the constants of integration.) How do your results depend on `N`?

12.8 Bilinear interpolation

Let $f(x, y)$ denote a function of independent variables x and y whose value is known only at discrete points in the x–y plane. We can use *bilinear interpolation* to estimate the values of f at other points in the plane.

Bilinear interpolation is just linear interpolation applied twice, once for each dimension. For simplicity let's consider a single "tile" in the x–y plane, bounded by $x_0 \leq x \leq x_1$ and $y_0 \leq y \leq y_1$. The function f is known only at the corners, (x_0, y_0), (x_0, y_1), (x_1, y_0), (x_1, y_1). Our interpolating function, which we will call $F(x, y)$, should match $f(x, y)$ at the corners of the tile. First, interpolate in the x-direction, from x_0 to x_1, along the line $y = y_0$:

$$F(x, y_0) = \frac{f(x_1, y_0) - f(x_0, y_0)}{x_1 - x_0}(x - x_0) + f(x_0, y_0). \tag{12.15}$$

This is just the linear interpolation formula (12.10) with some changes of notation. Next, we perform a linear interpolation along the line $y = y_1$:

$$F(x, y_1) = \frac{f(x_1, y_1) - f(x_0, y_1)}{x_1 - x_0}(x - x_0) + f(x_0, y_1). \tag{12.16}$$

Now we have estimates for the function values along the top and bottom edges of the tile. The final step is to interpolate in the y-direction, from y_0 to y_1, along the line $x = \text{const}$:

$$F(x, y) = \frac{F(x, y_1) - F(x, y_0)}{y_1 - y_0}(y - y_0) + F(x, y_0). \tag{12.17}$$

Combining these results, we find

$$F(x, y) = \frac{(x - x_1)(y - y_1)}{(x_1 - x_0)(y_1 - y_0)} f(x_0, y_0) - \frac{(x - x_0)(y - y_1)}{(x_1 - x_0)(y_1 - y_0)} f(x_1, y_0) - \frac{(x - x_1)(y - y_0)}{(x_1 - x_0)(y_1 - y_0)} f(x_0, y_1) + \frac{(x - x_0)(y - y_0)}{(x_1 - x_0)(y_1 - y_0)} f(x_1, y_1). \tag{12.18}$$

You can verify by inspection that F matches f at the corners of the tile. We would obtain the same result for $F(x, y)$ by interpolating first in the y-direction, then in the x-direction.

Exercise 12.8a

Create a function definition `F(x,y)` for the bilinear interpolating function of Eq. (12.18). Set $x_0 = y_0 = 0$ and $x_1 = y_1 = 1$, and use the data $f(0, 0) = -3.0$, $f(0, 1) = 2.0$, $f(1, 0) = 1.0$ and $f(1, 1) = 3.0$. Compare `F(0.5,0.5)` to the average of the corner values. Plot `F(x,y)` along the two diagonals, from (x_0, y_0) to (x_1, y_1) and from (x_0, y_1) to (x_1, y_0).

Exercise 12.8b

Create a contour plot of the interpolating function from the previous exercise. One way to do this: Set

```
xarray = np.linspace(x0,x1,N)
yarray = np.linspace(y0,y1,N)
Farray = np.zeros((N,N))
```

for some chosen `N`, then fill the array elements `Farray[i,j]` with `F(xarray[i],yarray[j])`. The command

```
matplotlib.pyplot.contour(Farray,[0])
```

will create the plot, where `[0]` is a list of contour values. Add more contours to your plot.

Chapter 13

Numerical Integration I

13.1 A simple example

Integrals cannot always be evaluated analytically. Here's an example. A self-driving car is programmed to travel along a straight road for one hour with velocity v (in kilometers per hour) as a function of time t (in hours) given by

$$v(t) = 100 \sin^2(\pi(t^2 - t)). \tag{13.1}$$

How far does the car travel during this one-hour trip? We know that velocity v is the time derivative of distance s, so the answer is found by integration:

$$s = \int_0^1 \left[100 \sin^2(\pi(t^2 - t))\right] dt. \tag{13.2}$$

This integral cannot be expressed analytically in terms of elementary functions. We must rely on numerical methods to find s.

Exercise 13.1

Write a Python code to plot the velocity $v(t)$ as a function of t for $0 \leq t \leq 1$. Note that the car accelerates smoothly from rest to a maximum of 50 km/hr, then decelerates smoothly and comes to a stop after 1 hour.

We will return to this problem later in this chapter, after presenting several techniques for evaluating integrals numerically.

13.2 Left endpoint rule

The integral

$$I = \int_a^b f(x)\,dx \tag{13.3}$$

is the "area under the curve" $f(x)$ from a to b.[1] The area can be approximated by the sum of the areas of rectangles, as shown in Fig. 13.1.

The rectangles are obtained by dividing the interval $a \le x \le b$ into $N+1$ nodes labeled x_0 through x_N. Note that x_0 is just another name for the left endpoint a. Likewise, x_N is another name for the right endpoint b. The $N+1$ nodes span N subintervals, labeled 1 through N. The first subinterval extends from x_0 to x_1. The Nth subinterval extends from x_{N-1} to x_N. In general, the ith subinterval extends from x_{i-1} to x_i.

Figure 13.1 depicts the *left endpoint rule* (also called the left Riemann sum). The height of the rectangle in the ith subinterval

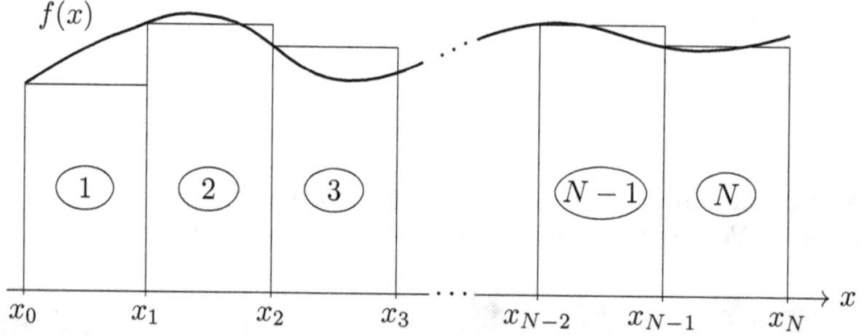

Fig. 13.1. The left endpoint rule (left Riemann sum) approximates the integral (13.3) as the sum of areas of rectangles. The figure shows the first three subintervals and the last two subintervals.

[1] This is the case for most functions we encounter in science and engineering. Some exotic functions require a more sophisticated definition of the integral.

is $f(x_{i-1})$, the value of the function at the left-side node. The area of the ith rectangle is the product of its height $f(x_{i-1})$ and its width $x_i - x_{i-1}$. We will take the nodes to be equally spaced, so that $x_i - x_{i-1} = \Delta x$ for each subinterval i. Then the left endpoint rule is

$$I_L = \sum_{i=1}^{N} f(x_{i-1})\Delta x. \tag{13.4}$$

The left endpoint rule only approximates the integral (13.3). You can see from Fig. 13.1 that I_L will not equal I exactly, since the rectangles are too tall in some regions and too short in other regions. In the limit as $N \to \infty$, the rectangles become infinitely thin and I_L approaches the exact value for the integral I.

Exercise 13.2a

A farmer plans to build a fence across one side of their property, which is 50 meters long. The fenceposts must be placed 1 meter apart. How many fenceposts does the farmer need?

Remember, when we approximate an integral, the number of nodes (fenceposts) is always one greater than the number of subintervals (space between fenceposts). If the number of subintervals is N, the number of nodes is $N+1$. Also observe that the sum in Eq. (13.4) is a sum over areas of rectangles. The rectangles fill the subintervals. Hence, the sum runs from 1 through N.

It is not difficult to implement the left endpoint rule (13.4) in a computer program. You need to specify the endpoints a, b and the number of subintervals N. You can define the integrand $f(x)$ using a function definition. The width of each subinterval is

```
Deltax = (b - a)/N
```

The node values x_i can be defined as a NumPy array using

```
x = np.linspace(a,b,N+1)
```

The sum can be carried out using a loop such as

```
for i in range(1,N+1):
```

Exercise 13.2b

Why do we divide $b - a$ by N, rather than $N + 1$, to obtain Δx? Why do we use $N + 1$ rather than N in the linspace command? Why do we use $N + 1$ rather than N in the range command?

Exercise 13.2c

Write a code to approximate $I = \int_a^b f(x)\,dx$ using the left endpoint rule. Test your code using $f(x) = \sin x$, $a = 0$ and $b = \pi/2$. What is the exact answer? Compare I_L to the exact answer for $N = 2$, 4, 8, 16, and 32.

13.3 Right endpoint and midpoint rules

For the *right endpoint rule* (also called the right Riemann sum), the height of each rectangle equals the function value at the right-side node. That is, for the ith subinterval, the rectangle height is $f(x_i)$. The right endpoint rule is

$$I_R = \sum_{i=1}^{N} f(x_i)\,\Delta x. \tag{13.5}$$

This approximates the integral I. In the limit as $N \to \infty$, the rectangles become infinitely thin and I_R approaches the exact value for I.

Exercise 13.3a

Write a code to approximate $I = \int_a^b f(x)\,dx$ using the right endpoint rule. Test your code using $f(x) = \sin x$, $a = 0$ and $b = \pi/2$. Compare I_R to the exact answer for $N = 2$, 4, 8, 16, and 32.

Consider a region where the function $f(x)$ is increasing. The left endpoint rule I_L underestimates the value of I, whereas the right endpoint rule overestimates the value of I. This is shown in Fig. 13.2.

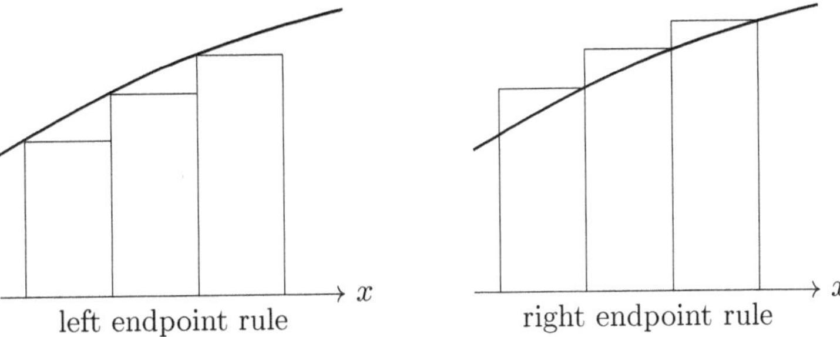

Fig. 13.2. For an increasing function $f(x)$, the left endpoint rule underestimates the area under the curve. The right endpoint rule overestimates the area under the curve.

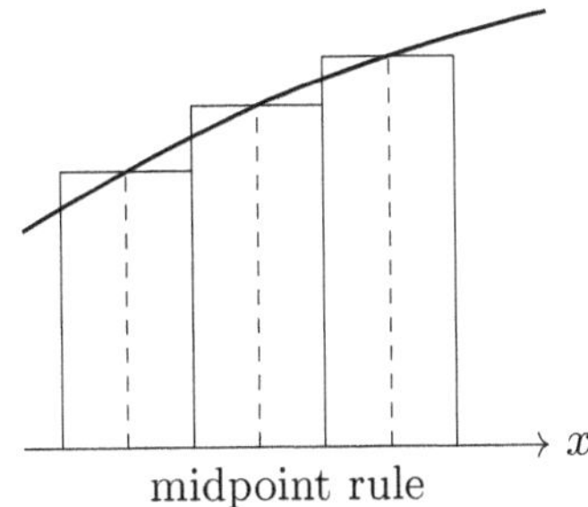

Fig. 13.3. For the midpoint rule, the height of each rectangle is the value of the function $f(x)$ at the midpoint of the subinterval.

Wouldn't it be better to evaluate the function at the midpoint of the subinterval? This should be more accurate since the extra area from the left half of the subinterval would tend to cancel the missing area from the right half of the subinterval.

This leads us to the *midpoint rule*, depicted in Fig. 13.3. The height of each rectangle is given by the value of the function at the midpoint of the subinterval. For the ith subinterval, the midpoint is $(x_{i-1}+x_i)/2$ and the height of the rectangle is $f((x_{i-1}+x_i)/2)$. The midpoint rule approximation to the integral I is

$$I_M = \sum_{i=1}^{N} f((x_{i-1}+x_i)/2)\,\Delta x. \tag{13.6}$$

In the limit as $N \to \infty$, the rectangles become infinitely thin and I_M approaches the exact value for I.

Exercise 13.3b

Write a code to approximate $I = \int_a^b f(x)\,dx$ using the midpoint rule. Test your code using $f(x) = \sin x$, $a = 0$ and $b = \pi/2$. Compare I_M to the exact answer for $N = 2$, 4, 8, 16, and 32. Compare the midpoint rule to the left and right endpoint rules.

13.4 Errors and convergence tests

The left endpoint rule I_L, right endpoint rule I_R, and midpoint rule I_M each approximate I with some error. In general, the size of the error will decrease if we increase the number of subintervals N. Increasing the number of subintervals N is equivalent to decreasing the size Δx of each subinterval, since $\Delta x = (b-a)/N$. "Decreasing the error" is also referred to as "increasing the resolution." We increase the resolution (or decrease the error) by increasing N.

A detailed mathematical analysis (see Ch. 14) shows that for smooth integrands, the error in the left and right endpoint rules is (approximately) inversely proportional to the number of subintervals. The error in the midpoint rule is (approximately) inversely proportional to the square of the number of subintervals. To summarize:

$$|\text{error for } I_L| \propto N^{-1},$$

$$|\text{error for } I_R| \propto N^{-1},$$

$$|\text{error for } I_M| \propto N^{-2}.$$

If we double the resolution for I_L or I_R, we reduce the error by (approximately) 1/2. If we double the resolution for I_M, we reduce the error by (approximately) 1/4.

For most numerical integration methods, the error scales like N^m where m is a *negative* integer. Note that $|\text{error}| \propto N^m$ is equivalent to

$$|\text{error}| = C\,N^m \tag{13.7}$$

for some constant C. Take the base 10 logarithm of this relation to obtain

$$\log|\text{error}| = m\log(N) + \log(C). \tag{13.8}$$

A plot of $\log|\text{error}|$ versus $\log(N)$ should be (approximately) a straight line with slope m.

It is always a good idea to carry out a *convergence test* of your numerical codes. Choose a smooth function whose integral is known analytically. Solve the integral numerically and determine the error as a function of resolution N. The error should scale as in Eq. (13.8); if not, there is a problem with your code.

Remember, in Python, the NumPy command `log` is the base e logarithm. The command for the base 10 logarithm is `log10`.

Exercise 13.4a

As in Exercise 13.2c, use the left endpoint rule to approximate the integral $\int_0^{\pi/2} \sin x\, dx$. Since the exact answer is known, you can carry out a convergence test: Compute the error at multiple resolutions. Copy and paste your results into a text file with N in the first column and the error in the second column. Create a code to read the text file and plot $\log|\text{error}|$ versus $\log(N)$. Does the curve approach a straight line with slope $m = -1$ as the resolution is increased?

Exercise 13.4b

Use the midpoint rule to approximate the integral $\int_0^{\pi/2} \sin x\, dx$ and perform a convergence test. Does the curve of $\log|\text{error}|$ versus $\log(N)$ approach a straight line with slope $m = -2$ as the resolution is increased?

These exercises show that as the resolution increases, the error in the midpoint rule decreases *more rapidly* than the error in the left endpoint rule. This has an important consequence. Let's say we want to know the answer for I to within a certain error, say, ± 0.0001. For most functions $f(x)$, the midpoint rule will require far fewer subintervals than the left or right endpoint rules to achieve a given level of accuracy.

Exercise 13.4c

Consider once again the integral $\int_0^{\pi/2} \sin x \, dx$. For the left endpoint rule, how many subintervals are required to obtain an error (in absolute value) of less than 0.0001? How many for the right endpoint rule? For the midpoint rule?

Exercise 13.4d

Write a code to compute

$$\int_1^{10} x \ln(10/x) \, dx$$

using the left endpoint and midpoint rules, where ln is the base e logarithm. What is the exact answer? With the left endpoint rule, approximately how many subintervals N are required to yield the answer with an error less than 0.01? Less than 0.001? With the midpoint rule, approximately how many subintervals N are required to yield the answer with an error less than 0.01? Less than 0.001?

13.5 Estimating errors and significant figures

In most cases you won't know the exact answer for a given integral, so there is no direct way of determining the error. For example, consider the integral

$$I = \int_0^{10} \sin(\pi \sin x) \, dx. \tag{13.9}$$

Using the midpoint rule with $N = 10$ subintervals, Python reports the result as $I_M = 1.3737955$. How accurate is this? Is it correct to 8 significant figures?

Before continuing, let's review the concept of *significant figures.* Roughly speaking, significant figures are numbers that are considered reliable. If a result such as 1.3737955 is only reliable to 3 significant figures, it should be written as 1.37. If it is reliable to 4 significant figures, it should be written as 1.374. (Note, we rounded up.)

Rules for significant figures:

- nonzero numbers are significant
- zeros are significant if
  - they appear between nonzero numbers
  - they are trailing numbers to the right of the decimal point
  - they are trailing numbers in a whole number that includes a decimal point.

For example the numbers 4030, 0.0250, and 640. all have 3 significant figures. The same rules apply to numbers written in scientific notation if we ignore the factor of 10 raised to some power. Thus, 3.050×10^7 has 4 significant figures.

With the midpoint rule and $N = 10$ subintervals, our code tells us that the integral (13.9) is $I_M = 1.3737955$. We would like to write this properly, with the correct number of significant figures, by taking into account the accuracy of the result.

Unfortunately we don't know how accurate the result is; equivalently, we don't know what the error is. We do know that the error is (approximately) proportional to N^{-2}, so we can reduce the error by increasing the resolution. If we double the resolution to $N = 20$, we find $I_M = 1.2664347$. This should be more accurate than 1.3737955. Perhaps the answer is around 1.26 or 1.27? To be sure, we need to increase the resolution even further. The following table shows the results obtained by repeatedly doubling the resolution:

N	I_L
10	1.3737955
20	1.2664347
40	1.2433224
80	1.2378149

It appears that the answers are settling down to about 1.23. Perhaps we should round up to 1.24? Note that the numbers are decreasing as we increase the resolution. Is the correct answer closer to 1.22?

At this point, we can be confident that the correct answer to 2 significant figures is 1.2. If we want a more accurate answer, we need to continue to higher resolution:

N	I_L
160	1.2364534
320	1.2361139
640	1.2360291

With these results, it is fairly clear that the answer is 1.236 to 4 significant figures. In fact, the answer appears to be converging to 1.2360, or perhaps 1.2359. If we want an answer that is accurate to 5 significant figures, we need to increase the resolution even further.

This example illustrates an important lesson: *You must run your code at multiple resolutions.* Until you do, you have no idea how accurate the answer is.

Exercise 13.5a

Use the midpoint rule to evaluate the integral of Sec. 13.1 for the distance traveled by the car during its one-hour trip. Show your results for multiple values of N, and use the data to determine the distance to 6 significant figures.

Exercise 13.5b

Use the midpoint rule to evaluate the integral

$$I = \int_{-10}^{10} \ln(2 + \sin(x))\,dx.$$

Make a table showing N and I_M for increasing values of N. Use your data to determine the value of the integral to 6 significant figures.

Chapter 14

Numerical Integration II

14.1 Takeaways from before

Numerical computation is all about efficiency. We want to find algorithms that give us the most accuracy with the least amount of computational effort.

In the previous chapter we discussed the left endpoint rule, right endpoint rule, and midpoint rule for numerically evaluating the integral

$$I = \int_a^b f(x)\,dx. \tag{14.1}$$

The errors in the left and right endpoint rules are proportional to N^{-1}, where N is the number of subintervals from a to b. To reduce the error by a factor of 100, you must increase N by a factor of 100. This requires 100 times as many evaluations of the integrand $f(x)$. This might be fine for simple problems, with simple integrands. But for complicated integrands that require a lot of computer time to evaluate, this can be a problem.

The midpoint rule is usually better than the left or right endpoint rules. The error for this method is proportional to N^{-2}. With the midpoint rule we can reduce the error by a factor of 100 by increasing N by a factor of 10. This requires only 10 times as many evaluations of the integrand $f(x)$.

Exercise 14.1

Consider the integral

$$I = \int_0^{\pi} 2x \sin(x^2)\, dx,$$

which has the exact value $I = 1 - \cos(\pi^2)$. Approximate this integral using the left endpoint rule I_L and midpoint rule I_M, both with $N = 10$ subintervals. Which rule gives the more accurate answer? How many subintervals are required to reduce the error in I_L by a factor of 100? How many subintervals are required to reduce the error in I_M by a factor of 100?

14.2 Error analysis

The following analysis shows why the error for the left endpoint rule is proportional to N^{-1}.

As always, divide the interval $a \le x \le b$ into N subintervals. Observe that the integral

$$\int_{x_{i-1}}^{x_i} dx\, f(x) \tag{14.2}$$

is the exact area under the curve $f(x)$ in the ith subinterval. Thus, we can write the integral I of Eq. (14.1) as

$$I = \sum_{i=1}^{N} \int_{x_{i-1}}^{x_i} f(x)\, dx, \tag{14.3}$$

which is a sum of integrals in each subinterval.

In the ith subinterval, expand the integrand $f(x)$ in a Taylor series about the left node:

$$f(x) = f(x_{i-1}) + f'(x_{i-1})(x - x_{i-1}) + \frac{1}{2} f''(x_{i-1})(x - x_{i-1})^2 + \cdots . \tag{14.4}$$

Here, $f'(x_{i-1})$ and $f''(x_{i-1})$ are the first and second derivatives of $f(x)$ evaluated at the left node x_{i-1}. Integrating over the subinterval, we find

$$\int_{x_{i-1}}^{x_i} f(x)\,dx = f(x_{i-1})\,\Delta x + \frac{1}{2} f'(x_{i-1})\,\Delta x^2 + \cdots . \tag{14.5}$$

Now sum this result over subintervals:

$$\sum_{i=1}^{N} \int_{x_{i-1}}^{x_i} f(x)\,dx = \sum_{i=1}^{N} f(x_{i-1})\,\Delta x + \sum_{i=1}^{N} \frac{1}{2} f'(x_{i-1})\,\Delta x^2 + \cdots . \tag{14.6}$$

The left-hand side is the exact integral I from Eq. (14.3), and the first term on the right-hand side is the left endpoint approximation I_L from Eq. (13.4). Thus, Eq. (14.6) can be written as

$$I = I_L + \sum_{i=1}^{N} \left\{ \frac{1}{2} f'(x_{i-1})\,\Delta x^2 + \cdots \right\}. \tag{14.7}$$

Therefore the error in the left endpoint rule, $I_L - I$, is

$$(\text{error for } I_L) = -\sum_{i=1}^{N} \left\{ \frac{1}{2} f'(x_{i-1})\,\Delta x^2 + \cdots \right\}. \tag{14.8}$$

The unwritten terms $(\cdots)$ are proportional to Δx^3 and higher powers of Δx.

Assuming f is smooth, so that the derivatives f', f'', etc. remain finite in the interval from a to b, the error terms will go to zero as Δx goes to zero (as the resolution is increased). The term that goes to zero least quickly is the one proportional to Δx^2. Thus,

$$(\text{error for } I_L) \approx -\frac{\Delta x}{2} \sum_{i=1}^{N} f'(x_{i-1})\,\Delta x, \tag{14.9}$$

where $\approx$ means "approximately equal."

The sum in Eq. (14.9) is actually the left endpoint rule approximation to the integral $\int_a^b f'(x)\,dx = f(b) - f(a)$. Therefore, the error becomes

$$(\text{error for } I_L) \approx \left[-\frac{1}{2} \left(f(b) - f(a) \right) \right] \Delta x. \tag{14.10}$$

We will use this result in the next section. For now, simply observe that the factor in square brackets is a constant, independent of Δx. Let's call this constant $C/(b-a)$. Since $\Delta x = (b-a)/N$, the error simplifies to

$$(\text{error for } I_L) \approx C\,N^{-1}. \tag{14.11}$$

That is, the error for the left endpoint rule is (approximately) proportional to N^{-1}.

A similar analysis shows that the error in the right endpoint rule is (approximately) proportional to N^{-1}.

Exercise 14.2a

The error analysis assumes the integrand is smooth. Compute

$$I = \int_0^9 \frac{1}{\sqrt{x}}\,dx$$

with the left and right endpoint rules. Note that the integrand is not smooth at $x = 0$. What is the exact answer? What result does the left endpoint rule give? Carry out a convergence test using the right endpoint rule. Is the error proportional to N^{-1}?

Exercise 14.2b

Show that the error in the midpoint rule is (approximately) proportional to N^{-2}. Hint: Expand $f(x)$ in a Taylor series about the midpoint $(x_i + x_{i-1})/2$. Integrate from x_{i-1} to x_i then sum over subintervals. You will need to argue that the factor $\sum_{i=1}^{N} f''((x_{i-1}+x_i)/2)\,\Delta x$ is a constant, independent of Δx.

14.3 Trapezoid rule

In the previous section we found an approximate expression, Eq. (14.10), for the error in the left endpoint rule. We can use this result to improve the left endpoint rule. Simply subtract the error from I_L to obtain the trapezoid rule,

$$I_T = I_L + \left[\frac{1}{2}\left(f(b) - f(a)\right)\right] \Delta x. \tag{14.12}$$

This numerical integration rule, like the midpoint rule, has errors proportional to N^{-2}. It is generally better than either the left or right endpoint rule.

Recall that $a = x_0$ and $b = x_N$. Using expression (13.4) for the left endpoint rule, we have

$$I_T = \sum_{i=1}^{N} f(x_{i-1})\,\Delta x + \frac{1}{2} f(x_N)\,\Delta x - \frac{1}{2} f(x_0)\,\Delta x. \tag{14.13}$$

Expand the sum,

$$\sum_{i=1}^{N} f(x_{i-1})\,\Delta x = \left[f(x_0) + f(x_1) + \cdots + f(x_{N-1})\right] \Delta x, \tag{14.14}$$

then rearrange terms to obtain

$$I_T = \left[\frac{1}{2} f(x_0) + f(x_1) + \cdots + f(x_{N-1}) + \frac{1}{2} f(x_N)\right] \Delta x. \tag{14.15}$$

This shows that the trapezoid rule can be written as

$$I_T = \frac{1}{2}\left[f(x_0) + f(x_N)\right] \Delta x + \sum_{i=1}^{N-1} f(x_i)\,\Delta x. \tag{14.16}$$

Note that the sum extends from 1 to $N - 1$.

Why is this numerical integration scheme called the trapezoid rule? Rearrange terms in Eq. (14.15) and write the trapezoid rule as follows:

$$\begin{aligned} I_T = &\frac{1}{2}\left[f(x_0) + f(x_1)\right] \Delta x + \frac{1}{2}\left[f(x_1) + f(x_2)\right] \Delta x + \cdots \\ &\cdots + \frac{1}{2}\left[f(x_{N-1} + f(x_N)\right] \Delta x. \end{aligned} \tag{14.17}$$

More compactly, we have

$$I_T = \sum_{i=1}^{N} \frac{1}{2} \left[f(x_{i-1}) + f(x_i) \right] \Delta x. \tag{14.18}$$

Written this way the trapezoid rule can be understood as a sum over subintervals. The area in each subinterval is approximated by the area of a trapezoid, as shown in Fig. 14.1. Recall that the area of a trapezoid is given by the average height, multiplied by the width. For the ith subinterval, the average height is the average of the function values at each end node, namely, $\left[f(x_{i-1}) + f(x_i) \right]/2$.

Exercise 14.3a

Write a code that uses the trapezoid rule, Eq. (14.16), to approximate

$$I = \int_1^2 \sin(x + \ln x)\, dx.$$

Make a table showing the answer for various numbers of subintervals N. Determine the answer to at least 5 significant figures.

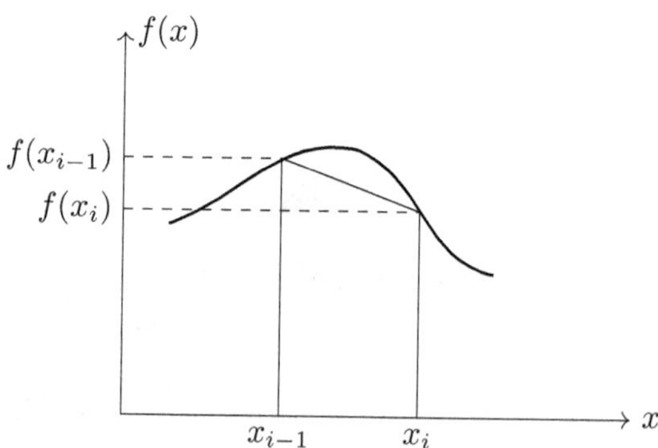

Fig. 14.1. The trapezoid rule. The area under the curve in the ith subinterval is approximated by the area of a trapezoid.

Exercise 14.3b

Test your trapezoid rule code with

$$I = \int_1^{10} x \ln(x)\, dx.$$

What is the exact solution? Carry out a convergence test by plotting log |error| versus resolution $\log(N)$. Do the errors scale scale like N^{-2}?

14.4 Simpson's rule

For numerical integration, the computer spends most of its time evaluating the integrand $f(x)$ at various values of x. Look closely at the formulas for the left endpoint rule (13.4), right endpoint rule (13.5), midpoint rule (13.6) and trapezoid rule (14.16). For the left and right endpoint rules, the function $f(x)$ must be evaluated N times. Likewise, for the midpoint rule, the integrand $f(x)$ must be evaluated N times. The trapezoid rule is similar, requiring $N + 1$ evaluations of the integrand.

Although these four methods require about the same number of function evaluations, the midpoint rule I_M and trapezoid rule I_T are generally much better than the left and right endpoint rules because their errors decrease more rapidly with increasing resolution:

$$|\text{errors for } I_L \text{ and } I_R| \propto N^{-1},$$
$$|\text{errors for } I_M \text{ and } I_T| \propto N^{-2}.$$

Can we do even better? Can we find an algorithm whose errors decrease even more rapidly than N^{-2} without a drastic increase in the number of evaluations of the integrand $f(x)$? The answer is yes—Simpson's rule is such an algorithm.

Simpson's rule is defined by

$$I_S = \frac{\Delta x}{6}\left[f(x_0) + f(x_N) + 2\sum_{i=1}^{N-1} f(x_i) + 4\sum_{i=1}^{N} f((x_{i-1} + x_i)/2)\right]. \tag{14.19}$$

Geometrically, Simpson's rule is obtained by approximating the area in each subinterval as the area under a parabola that matches the function $f(x)$ at each endpoint and at the midpoint.

For Simpson's rule the integrand must be evaluated at each node and at the midpoint of each subinterval. For N subintervals, this is $2N+1$ function evaluations. That's roughly twice the number of evaluations as the previous integration rules. However, the error in Simpson's rule decreases very rapidly with increasing resolution:

$$|\text{errors for } I_S| \propto N^{-4}.$$

In most cases, this decrease more than compensates for the increase in the number of function evaluations.

Exercise 14.4a

Use Simpson's rule to approximate

$$I = -\int_{\pi^2/4}^{4\pi^2} \sin(\sqrt{x})\, dx.$$

The exact answer is $I = 2 + 4\pi$. Find the number of subintervals required for Simpson's rule to compute the answer to 8 significant figures. How many evaluations of the integrand does this require? Approximate the same integral using the trapezoid rule, and find the number of subintervals required to reach 8 significant figures. How many evaluations of the integrand are needed for the trapezoid rule?

Exercise 14.4b

Approximate the integral

$$I = \int_0^{\pi} (\sin x)^{3/2}\, dx$$

using Simpson's rule. Find the answer to 6 significant figures.

Exercise 14.4c

Consider the integral

$$I = \int_0^{2\pi} x^2 \cos x \, dx.$$

Show that the exact answer is 4π. Compute the error in Simpson's rule for various numbers of subintervals N. Perform a convergence test to show that the error is proportional to N^{-4}.

14.5 Integration with SciPy

The Python library SciPy has a number of built-in functions for integration. For example, the following code uses the SciPy function `quad()` to evaluate the integral $I = \int_0^{\pi/2} \sin x \, dx$:

```
import scipy.integrate as si
import numpy as np

def f(x):                # Define the  integrand
    return np.sin(x)

answer, error = si.quad(f,0,np.pi/2)
print(answer, error)
```

The exact answer is 1. The function `quad()` returns the (approximate) answer 0.9999999999999999 along with an error estimate of $1.1102230246251564e - 14$.

Exercise 14.5a

The Fresnel integrals defined by

$$S(u) = \int_0^u \sin(\pi x^2/2) dx,$$

$$C(u) = \int_0^u \cos(\pi x^2/2) dx,$$

play an important role in optics. Write a code to create a plot of $S(u)$ and $C(u)$ versus u for $0 \leq u \leq 5$. Use `quad()` to evaluate the integrals.

The integrand might depend on parameters as well as the integration variable. This code computes $\int_0^{\pi/2} a\sin(bx)\,dx$ with $a = 2.0$ and $b = \pi$:

```
def f(x,a,b):                # Define integrand
        return a*np.sin(b*x)

answer, error = si.quad(f,0,np.pi/2,(2.0,np.pi))
print(answer, error)
```

The function `quad()` passes the values `2.0` and `np.pi` to `f()` as the Python "tuple" `(2.0,np.pi)`. The elements of a tuple are enclosed in parentheses. The function `f()` assigns these values to `a` and `b`.

Exercise 14.5b

The Bessel functions of the first kind, defined by

$$J_n(x) = \frac{1}{\pi}\int_0^{\pi} \cos(x\sin t - nt)\,dt,$$

appear in many contexts such as the vibrations of a drumhead. Write a code that plots $J_n(x)$ versus x in the domain $0 \leq x \leq 20$, for $n = 0, 1, 2, 3$. Use `quad()` to evaluate the integrals. (Note that the integration variable is t.)

Exercise 14.5c

Euler's constant

$$\gamma = \int_1^{\infty} \left(\frac{1}{\lfloor x \rfloor} - \frac{1}{x}\right) dx$$

plays an important role in mathematics. Here, $\lfloor x \rfloor$ is the floor function, the integer part of x. NumPy has a built-in floor function called `floor()`. The integral definition of Euler's constant is difficult to evaluate because the integrand is discontinuous at integer values of x, and because the range of integration is infinite. Write a code to evaluate γ using the following strategy. Break the range of integration into subintervals from 1 to 2, from 2 to 3, etc. Use `quad()` to evaluate the integral in each subinterval and add the results. Keep adding subintervals until the answer settles down to 4 significant figures.

Exercise 14.5d

The gamma function is defined by

$$\Gamma(s) = \int_0^\infty x^{s-1} e^{-x}\, dx.$$

Since the integrand approaches zero rapidly as x increases, you can approximate the integral by replacing the upper limit of integration with a large positive constant u and using `quad()`. The gamma function has the amazing property that for positive integer values of its argument, $\Gamma(s) = (s-1)!$. Use this as a check to make sure your choice for u is large enough to give reasonably accurate answers. Plot a graph of $\Gamma(s)$ for the domain $0.2 \le s \le 4$.

14.6 Nonuniform data

The `scipy.integrate` function `simps()` can be used to integrate a function whose values are only known at discrete points. This situation can arise when data are obtained from an experiment. For example, let's assume that only the following function values are known:

$$f(1) = 7$$

$$f(2) = 5$$

$$f(4) = 12$$

$$f(5) = 10$$

$$f(8) = 6$$

This code computes the integral of $f(x)$ from $x = 1$ to $x = 8$:

```
x = np.array([1,2,4,5,8])
y = np.array([7,5,12,10,6])

answer = si.simps(y,x)
print(answer)
```

Exercise 14.6

Write a code that will read the data *NonUniformData.txt*, generated by the code given in the Appendix. The data consist of two columns, x and $f(x)$. Have your code plot the data and compute the integral of $f(x)$.

14.7 Monte Carlo integration

In computational science, the term “Monte Carlo” refers to a host of numerical techniques that rely on random sampling to solve a problem. In particular, Monte Carlo methods can be used to evaluate integrals numerically.

Take another look at the midpoint rule, Eq. (13.6). Since $\Delta x = (b - a)/N$, we can rewrite this as

$$I_M = \left[\frac{1}{N} \sum_{i=1}^{N} f((x_{i-1} + x_i)/2) \right] (b - a). \tag{14.20}$$

The factor in square brackets is the average value of the function $f(x)$ in the interval from a to b. Let $\langle f \rangle$ denote this average; that is,

$$\langle f \rangle = \frac{1}{N} \sum_{i=1}^{N} f((x_{i-1} + x_i)/2). \tag{14.21}$$

In this case the average is obtained by evaluating $f(x)$ at N equally spaced points from a to b. As an alternative, we can compute the average $\langle f \rangle$ using N randomly selected points from a to b. Let $\mathcal{X}$ denote a set of N random numbers selected from the integration region $a \le x \le b$; then

$$\langle f \rangle = \frac{1}{N} \sum_{x \in \mathcal{X}} f(x). \tag{14.22}$$

The Monte Carlo approximation to $\int_a^b f(x)\,dx$ is simply $\langle f \rangle (b - a)$, or

$$I_{MC} = \frac{(b-a)}{N} \sum_{x \in \mathcal{X}} f(x). \tag{14.23}$$

In the limit as $N \to \infty$, the Monte Carlo approximation I_{MC} approaches the exact value for the integral.

To implement the Monte Carlo method, use NumPy's default random number generator with the function `uniform(`*low*,*high*`)` discussed in Sec. 8.3. Each time this function is called it creates a random number between *low* (inclusive) and *high* (exclusive).[1] Alternatively, the command `uniform(`*low*,*high*,*N*`)` will produce a NumPy array of N random real numbers between *low* and *high*.

Exercise 14.7

Use the Monte Carlo method to evaluate the Fresnel integral $S(u)$ at $u = 10$. Each time you run your code, the answer will be a bit different. How large must N be so that the result is usually accurate to about 2 significant figures?

---

[1]For real numbers, we don't need to worry that *low* is inclusive but *high* is exclusive. After all, the computer can only approximate most real numbers to machine accuracy, as discussed in Sec. 4.4.

14.8 Multidimensional integrals

Monte Carlo is not very practical for one-dimensional integrals. It usually takes a lot of random points N to obtain a modest degree of accuracy. Monte Carlo methods are primarily used to evaluate integrals over higher-dimensional regions with complex shapes. For such problems, Monte Carlo can be relatively simple and efficient.

Here's an example. Evaluate the integral

$$I = \int_{\mathcal{R}} \exp(x + y)\, dx\, dy, \tag{14.24}$$

where $\mathcal{R}$ is the interior of the ellipse

$$\frac{x^2}{4} + y^2 = 1. \tag{14.25}$$

The Monte Carlo approximation to I is given by $I_{\mathrm{MC}} = A_{\mathrm{ell}}\langle f \rangle$, where A_{ell} is the area of the ellipse and $\langle f \rangle$ is the average value of the integrand $f(x, y) = \exp(x + y)$ within the ellipse.

In Fig. 14.2, we see that the ellipse is bounded by the rectangle with $-2 \le x \le 2$ and $-1 \le y \le 1$. We can find the average value of the integrand within the ellipse by choosing random points (x, y) inside the rectangle, and discarding the points that lie outside the ellipse. The area of the ellipse can be found by computing the fraction

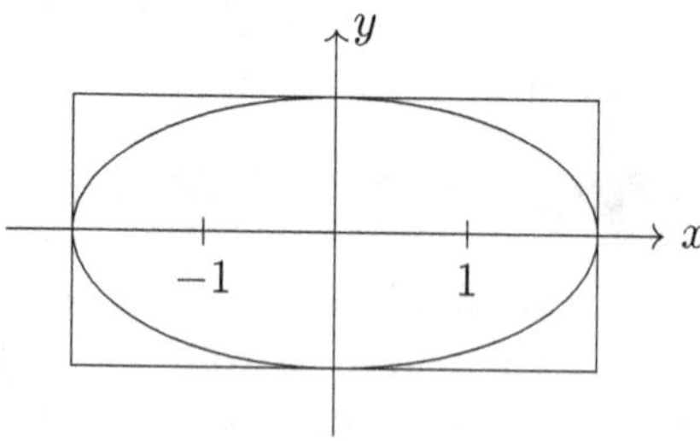

Fig. 14.2. The ellipse $x^2/4 + y^2 = 1$ is contained within the rectangle $-2 \le x \le 2$, $-1 \le y \le 1$.

of random points that lie inside the ellipse, then multiplying this fraction by the area of the rectangle. Here is a complete code:

```
import numpy as np

def f(x,y):                        # define the integrand
    return np.exp(x + y)

rnum = np.random.default_rng() # random number generator

Ntot = 200000         # total number of points
sumf = 0.0            # sum of function values
Nin = 0               # number of points inside ellipse

for n in range(Ntot):              # loop over all points
    x = rnum.uniform(-2,2)         # random x value
    y = rnum.uniform(-1,1)         # random y value
    if x**2/4.0 + y**2 > 1:        # (x,y) is outside ellipse
        continue                   # discard that point
    else:                          # (x,y) is inside ellipse
        Nin = Nin + 1              # increment inside points
        sumf = sumf + f(x,y)       # add f to sum

arearec = 8                        # area of rectangle
areaell = (Nin/Ntot)*arearec       # area of ellipse
averagef = sumf/Nin                # average value of f
IMC = areaell*averagef             # answer
print(areaellipse, IMC)
```

Exercise 14.8

Use the Monte Carlo method to evaluate the integral

$$\int_{\mathcal{R}} dx\, dy\, \ln(1 + xy),$$

where $\mathcal{R}$ is the region defined by the unit square in the first quadrant with the circle $\sqrt{(x - 1/2)^2 + (y - 1/2)^2} = 1/3$ removed. (See Fig. 14.3.) Use enough points N to deduce the correct answer to three significant figures.

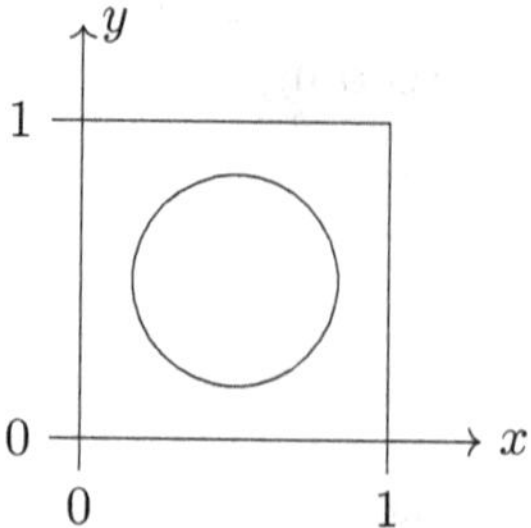

Fig. 14.3. The region of integration for Exercise 14.8 lies between the circle and the square.

14.9 More Exercises

Exercise 14.9a

The *error function* is a special function defined by

$$\operatorname{erf}(x) \equiv \frac{2}{\sqrt{\pi}} \int_0^x e^{-t^2}\, dt.$$

Write a code to create a plot of $\operatorname{erf}(x)$ versus x for $-3 \leq x \leq 3$. Use the SciPy function `quad()` to evaluate the integral.

Exercise 14.9b

Write a function definition for

$$f(x) = -1 + \int_1^x \frac{1}{t} dt,$$

where Simpson's rule is used to evaluate the integral. Apply the bisection method to find the root of this function. This is one way to determine e, the base of the natural logarithm.

Chapter 15

Linear Algebra

Linear algebra is one of the most important subjects in mathematics, with applications found in every branch of science. The techniques of linear algebra also form the basis of many numerical algorithms. In this chapter, we will translate linear algebra problems into matrix notation and solve them using NumPy commands.

15.1 Matrix multiplication

Here is a quick review of matrix multiplication. Consider the 2×2 matrices A and B defined by

$$A = \begin{pmatrix} 1 & 2 \\ 3 & 4 \end{pmatrix}, \quad B = \begin{pmatrix} 5 & 6 \\ 7 & 8 \end{pmatrix}.$$

The following equation depicts the calculation of the product AB:

$$\begin{pmatrix} \boxed{1 \ \ 2} \\ 3 \ \ 4 \end{pmatrix} \begin{pmatrix} \boxed{\begin{matrix} 5 \\ 7 \end{matrix}} & \begin{matrix} 6 \\ 8 \end{matrix} \end{pmatrix} = \begin{pmatrix} \boxed{1 \cdot 5 + 2 \cdot 7} & 1 \cdot 6 + 2 \cdot 8 \\ 3 \cdot 5 + 4 \cdot 7 & 3 \cdot 6 + 4 \cdot 8 \end{pmatrix} = \begin{pmatrix} \boxed{19} & 22 \\ 43 & 50 \end{pmatrix}.$$

The entry in the first row and first column of AB is the dot product of the first row of A and the first column of B. Likewise, the entry in the first row and second column of AB is the dot product of the first row of A and the second column of B. In general, the entry in the nth row and mth column of AB is the dot product of the nth row of A and the mth column of B. The same pattern holds for multiplication of any two matrices. The only restriction is that the number of entries in the rows of A must match the number of entries in the columns of B.

15.2 Linear systems

A basic problem in linear algebra is the solution of n linear algebraic equations for n unknowns. Consider the two equations

$$x + 3y = -3, \tag{15.1a}$$

$$2x - 2y = 10, \tag{15.1b}$$

for the two unknowns x and y. You can verify that the solution is $x = 3$, $y = -2$.

The problem above can be written in matrix notation as

$$\begin{pmatrix} 1 & 3 \\ 2 & -2 \end{pmatrix} \begin{pmatrix} x \\ y \end{pmatrix} = \begin{pmatrix} -3 \\ 10 \end{pmatrix}. \tag{15.2}$$

More compactly, we write this as $Av = r$ where

$$A = \begin{pmatrix} 1 & 3 \\ 2 & -2 \end{pmatrix}, \quad v = \begin{pmatrix} x \\ y \end{pmatrix}, \quad r = \begin{pmatrix} -3 \\ 10 \end{pmatrix}. \tag{15.3}$$

A is called the coefficient matrix. The column vector v contains the unknowns x and y. The column vector r is the right-hand side of the equation.

The problem of solving $Av = r$ for the unknowns v can now be viewed as the problem of inverting the matrix A. Let A^{-1} denote the inverse of A.

Exercise 15.2a

Verify that

$$A^{-1} = \begin{pmatrix} 1/4 & 3/8 \\ 1/4 & -1/8 \end{pmatrix}$$

is the inverse of A. That is, use matrix multiplication to show that $A^{-1}A = I$ and $AA^{-1} = I$, where

$$I = \begin{pmatrix} 1 & 0 \\ 0 & 1 \end{pmatrix}$$

is the identity matrix.

Having found the inverse of A, we can now solve our problem. Multiply the equation $Av = r$ by A^{-1} to obtain $Iv = A^{-1}r$. You can verify that $Iv = v$; that's why I is called the "identity matrix." Then the solution for the unknowns x and y is $v = A^{-1}r$. Explicitly, this is

$$\begin{pmatrix} x \\ y \end{pmatrix} = \begin{pmatrix} 1/4 & 3/8 \\ 1/4 & -1/8 \end{pmatrix} \begin{pmatrix} -3 \\ 10 \end{pmatrix}. \tag{15.4}$$

Carrying out the multiplication, we find

$$\begin{pmatrix} x \\ y \end{pmatrix} = \begin{pmatrix} 3 \\ -2 \end{pmatrix}. \tag{15.5}$$

Thus, $x = 3$ and $y = -2$.

This problem is easily solved using the `matrix` data type defined by NumPy. Let's assume NumPy is loaded as `np`. The matrix A is defined by

```
A = np.matrix([[1,3],[2,-2]])
```

Note the syntax. In general, NumPy matrices are defined by

`np.matrix`([[*first row*] , [*second row*] , [*third row*]])

The argument of `matrix()` is contained in round parentheses (). The matrix itself uses two sets of nested square brackets. The outer set of brackets contains the list of rows. Each inner set of brackets contains a list of elements for a row.

The inverse A^{-1} can be computed using

```
Ainv = np.linalg.inv(A)
```

The function `inv()` is contained in the linear algebra subpackage `linalg` within NumPy. The inverse of A can also be computed using the abbreviated notation `Ainv = A.I`.

Now define the right-hand side:

```
r = np.matrix([[-3],[10]])
```

We might refer to r as a column vector, but note that it is defined as a NumPy matrix with two rows and one column. What's the difference between `np.matrix([[-3,10]])` and `np.matrix([[-3],[10]])`?

Matrix multiplication is denoted by an asterisk $*$. Thus, the answer to our problem, $v = A^{-1}r$, is computed as

```
v = Ainv*r
print(v)
```

This yields the expected result,

```
matrix([[3],
        [-2]])
```

That is, $x = 3$ and $y = -2$.

Although the solution can be viewed as a column vector, $v = \binom{3}{-2}$, the calculation `v = Ainv*r` produces a NumPy matrix with two rows and one column. Thus, if we want to access the first element of v, which is the value of x, we must type `v[0,0]`. The second element, the value of y, is `v[1,0]`.

Exercise 15.2b

Translate the linear system

$$-2x + 7y + z = 3$$

$$4x - 5y + 3z = -2$$

$$2x - y + 4z = 6$$

into the form $Av = r$. Solve for x, y, and z by inverting the coefficient matrix A and computing $v = A^{-1}r$.

15.3 The `linalg.solve()` function

In general, solving a system of linear equations by inverting the coefficient matrix is a bad idea. Finding the numerical inverse of a matrix is:

- Time consuming. Other methods of solving linear systems are faster.
- Inaccurate. Inverting a matrix can lead to large machine roundoff errors.

As an example of this second point, consider the matrix

$$A = \begin{pmatrix} 4001 & 2001 \\ 8000 & 4001 \end{pmatrix}. \tag{15.6}$$

This is an "ill-conditioned" matrix that is difficult to invert numerically.

Exercise 15.3a

Use `linalg.inv()` to compute the inverse of the matrix A from Eq. (15.6), then compute $A^{-1}A$. Compare this to the identity matrix. How large are the errors? Carry out the same calculation for the matrix

$$B = \begin{pmatrix} 3001 & 2001 \\ 8000 & 4001 \end{pmatrix},$$

which is not ill-conditioned. How large are the errors in $B^{-1}B$?

Fortunately, linear algebra is a mature subject and the numerical routines available in the `numpy.linalg` library are well developed and reliable. In particular, we can use the `solve()` function to solve the linear system $Av = r$:

```
v = np.linalg.solve(A,r)
```

The algorithm used by `solve()` can be faster and more accurate than explicit matrix inversion.

Exercise 15.3b

Solve the linear system

$$\begin{aligned} 1.2x - 3.6y + 4.5z &= 1.3, \\ -3.3x + 4.2y - 8.1z &= -2.5, \\ -0.9x - 3.1y + 0.9z &= 0.1, \end{aligned}$$

in two different ways. First, use `linalg.inv()` to find the inverse of the coefficient matrix A, and compute $v = A^{-1}r$. Second, use `linalg.solve()`. In each case, have your code check the result by comparing $Av - r$ to the zero vector.

From now on, you should use `solve()` rather than matrix inversion to solve linear systems of equations.

15.4 Statics

A rigid, uniform beam of length ℓ and mass m is connected to a wall at one end and supported by a cord attached to the other end. See Fig. 15.1. The cord makes an angle θ with the beam, and has tension T. The wall exerts a force on the beam with components F_x, F_y. Since the beam is in static equilibrium Newton's laws of motion tell us that the sum of forces must vanish, $\sum \vec{F} = 0$. The forces lie in the x–y plane (the plane of the diagram), so $\sum \vec{F} = 0$ yields two equations:

$$F_x - T\cos\theta = 0, \tag{15.8a}$$

$$F_y + T\sin\theta - mg = 0. \tag{15.8b}$$

Likewise, the sum of torques about any point $\mathcal{P}$ must vanish. This reduces to a single equation for the torque about the z-axis,

$$(T\sin\theta)(\ell/2) - F_y(\ell/2) = 0, \tag{15.9}$$

where we have used the middle of the beam as the point $\mathcal{P}$.

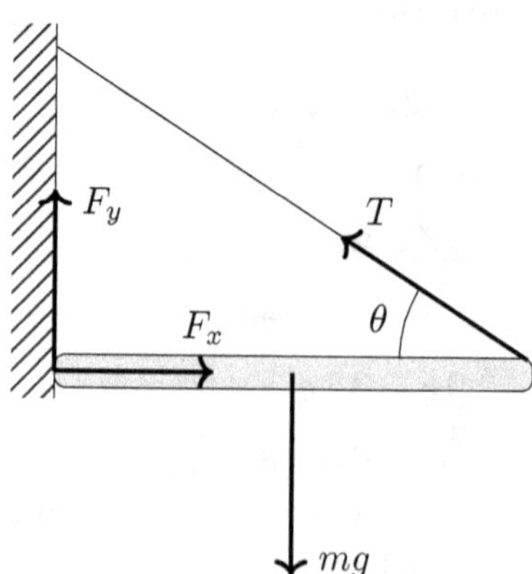

Fig. 15.1. A static beam of mass m and length ℓ is connected to a wall and a cord.

These three equations can be written in matrix form $Av = r$:

$$\begin{pmatrix} 1 & 0 & -\cos\theta \\ 0 & 1 & \sin\theta \\ 0 & -\ell/2 & \ell\sin\theta/2 \end{pmatrix} \begin{pmatrix} F_x \\ F_y \\ T \end{pmatrix} = \begin{pmatrix} 0 \\ mg \\ 0 \end{pmatrix}. \tag{15.10}$$

Assuming mg, ℓ, and θ are known, we can solve this equation for the unknown forces F_x, F_y, and T.

Exercise 15.4a

Solve Eq. (15.10) numerically for a beam of mass $m = 3.5\,\text{kg}$ and length $\ell = 1.2\,\text{m}$, and an array of cord angles θ. (The acceleration due to gravity is $g = 9.8\,\text{m/s}^2$.) Have your code plot a graph of tension T versus angle θ.

Exercise 15.4b

The beam in Fig. 15.2 has length ℓ, mass m, and makes an angle θ with respect to the floor. The cord is horizontal with tension T. Derive the equations of static equilibrium for the beam. Write your equations in matrix form $Av = r$ where the unknowns are the tension in the cord and the components of force that the floor exerts on the beam. Solve the matrix equation numerically using reasonable values for mg, ℓ, and θ.

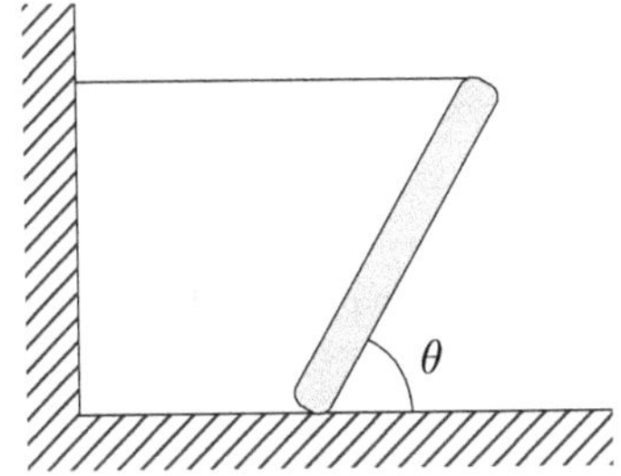

Fig. 15.2. A static beam with one end resting on the floor, the other end connected to a horizontal cord. The force that the floor exerts on the beam, the downward force due to gravity, and the tension in the cord are not shown.

15.5 Kirchoff's laws

Kirchoff's laws describe electrical circuits with steady currents. Figure 15.3 shows a simple electrical circuit consisting of two batteries, $\mathcal{E}_1$ and $\mathcal{E}_2$, and three resistors, R_1, R_2, and R_3. Currents I_1, I_2, and I_3 flow through the three sections of the circuit.

Kirchoff's first law states that the current entering a node is equal to the current exiting a node. At node a, the current entering is I_1+I_2 and the current exiting is I_3; therefore,

$$I_1 + I_2 = I_3. \tag{15.11}$$

Kirchoff's first law applied to node b yields the same relation.

Kirchoff's second law states that the sum of voltages around any closed loop must vanish. Recall that when a current I passes through a resistor R, the voltage drops by an amount IR. Kirchoff's second law applied to loop 1 of Fig. 15.3 gives

$$\mathcal{E}_1 - I_1 R_1 - I_3 R_3 = 0. \tag{15.12a}$$

For loop 2, we have

$$\mathcal{E}_2 - I_2 R_2 - I_3 R_3 = 0. \tag{15.12b}$$

We can apply Kirchoff's second law to the large loop around the outer edge of the circuit. The result is just a linear combination of the previous two equations.

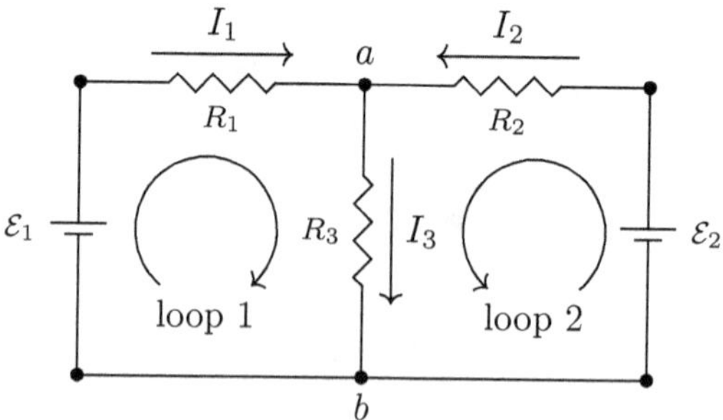

Fig. 15.3. A simple circuit with two batteries and three resistors. Letters a and b label nodes.

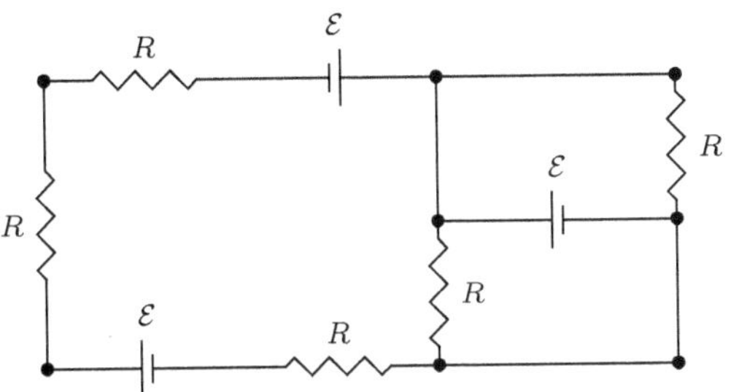

Fig. 15.4. Circuit for Exercise 15.5b.

Exercise 15.5a

Express Eqs. (15.11), (15.12) in matrix form, with the currents I_1, I_2, and I_3 as unknowns. Write a code to solve for the currents using the data $\mathcal{E}_1 = 12\,\text{V}$, $\mathcal{E}_2 = 9\,\text{V}$, $R_1 = 100\,\Omega$, $R_2 = 120\,\Omega$, $R_3 = 65\,\Omega$.

Exercise 15.5b

Use Kirchoff's laws to analyze the circuit in Fig. 15.4. The five resistors are identical, and the three batteries are identical. Set $R = 100\,\Omega$ and $\mathcal{E} = 9\,\text{V}$ and solve numerically for the current in each part of the circuit. (Hint: there are six currents. Use three Kirchoff first laws and three Kirchoff second laws.)

15.6 Eigenvalues and eigenvectors

Another important class of problems in linear algebra is the eigenvalue–eigenvector problem. Given a matrix M, the goal of this problem is to find numbers λ and column vectors v such that

$$Mv = \lambda v. \tag{15.13}$$

The numbers λ are the *eigenvalues*, and v are the *eigenvectors*.

The eigenvalue–eigenvector equation can be written as $Mv = \lambda Iv$, where I is the identity matrix. Bringing both terms to the left-hand side, we have $(M - \lambda I)v = 0$. This equation has the form $Av = 0$, where the matrix A is defined by $A \equiv M - \lambda I$. We already know how to solve such a system: simply invert the matrix A and multiply both sides by A^{-1}. (Better yet, use the `linalg.solve()` function in NumPy.) This calculation yields the trivial solution $v = A^{-1}0 = 0$. Of course $v = 0$ is always a solution to our original problem, $Mv = \lambda v$. In fact, $v = 0$ is the only solution if the matrix $A \equiv M - \lambda I$ is invertible. But for certain special values of λ, the matrix $M - \lambda I$ will *not* be invertible. In those cases, the system $Mv = \lambda v$ will have nontrivial solutions for the eigenvectors v.

Our first step in solving the eigenvalue–eigenvector problem consists in finding the eigenvalues λ. That is, we look for values of λ such that $M - \lambda I$ is not invertible.[1] Then for each eigenvalue we find a corresponding eigenvector v that satisfies $Mv = \lambda v$. Typically, for an $n \times n$ matrix M, the eigenvalue–eigenvector equation will have n solutions. That is, there will be n eigenvalues λ, and for each eigenvalue there will be a corresponding eigenvector v.

In Python, you can use the NumPy function `linalg.eig()` to solve the eigenvalue–eigenvector problem. More precisely, the command

```
eval,evec = np.linalg.eig(M)
```

will compute the eigenvalues and corresponding eigenvectors of a matrix M. The eigenvalues are placed into a NumPy array, called `eval` in this example. The eigenvectors are placed into a NumPy matrix, called `evec` in this example. The first column, `evec[:,0]`, is the eigenvector corresponding to the first eigenvalue, `eval[0]`; the second column, `evec[:,1]`, is the eigenvector corresponding to the second eigenvalue, `eval[1]`; etc. A colon : in place of a matrix index means "all elements."

Exercise 15.6

Write a code that uses `linalg.eig()` to solve the eigenvalue–eigenvector problem for

$$M = \begin{pmatrix} 1 & 3 \\ 4 & 2 \end{pmatrix}.$$

There are two eigenvalues and two corresponding eigenvectors. Verify that $Mv = \lambda v$ is satisfied for each.

[1]A matrix A is not invertible if its determinant $\det(A)$ vanishes. Thus, the eigenvalues satisfy $\det(M - \lambda I) = 0$.

Note that the eigenvector associated with a given eigenvalue is not unique. If v satisfies the equation $Mv = \lambda v$, then so does any vector proportional to v. The eigenvectors returned by the `linalg.eig()` function are normalized so that the sum of the squares of the components is unity.

15.7 Normal mode analysis

Two masses move in one dimension along a frictionless table, as shown in Fig. 15.5. The masses are connected between fixed walls by three springs. The separation between walls is D. Each spring has stiffness k and relaxed length ℓ_0. The force that a spring exerts on the objects attached to its ends is $\pm k(\ell - \ell_0)$, where ℓ is the length of the spring.

Let X_L denote the position of the mass on the left and X_R denote the position of the mass on the right. The lengths of the three springs are (respectively, from left to right) X_L, $X_R - X_L$ and $D - X_R$. The first spring exerts a force $-k(X_L - \ell_0)$ on the left mass. The second spring exerts a force $+k(X_R - X_L - \ell_0)$ on the left mass and a force $-k(X_R - X_L - \ell_0)$ on the right mass. The third spring exerts a force $+k(D - X_R - \ell_0)$ on the right mass. Newton's second law for the two masses yields

$$m\ddot{X}_L = -k(X_L - \ell_0) + k(X_R - X_L - \ell_0), \tag{15.14a}$$

$$m\ddot{X}_R = -k(X_R - X_L - \ell_0) + k(D - X_R - \ell_0), \tag{15.14b}$$

where dots denote time derivatives.

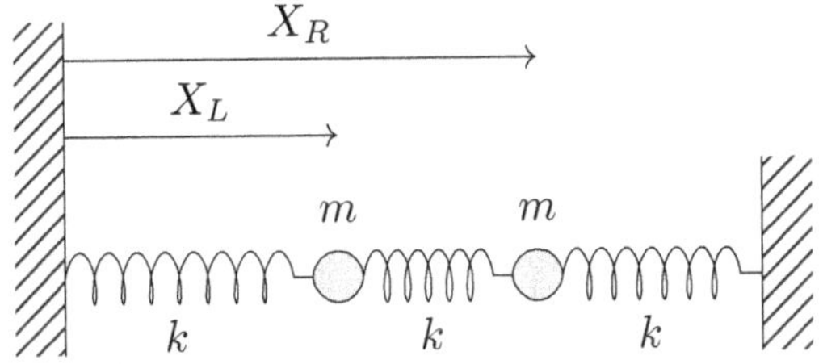

Fig. 15.5. Two masses connected to walls by three springs. The distance between walls is D.

The *normal mode* analysis is a powerful technique used to characterize the oscillations of a system near equilibrium. A normal mode is a pattern of oscillation in which each part of the system vibrates about its equilibrium position with a common angular frequency ω.

The equilibrium configuration for the system is found by setting $\ddot{X}_L = \ddot{X}_R = 0$ in Eqs. (15.14) and solving for X_L and X_R.

Exercise 15.7a

Verify that the equilibrium solution is $X_L = D/3$, $X_R = 2D/3$.

Let $x_L = X_L - D/3$ denote the position of the left mass relative to its equilibrium position. Likewise, $x_R = X_R - 2D/3$ denotes the position of the right mass relative to its equilibrium position. In terms of these new coordinates, the equations of motion become

$$m\ddot{x}_L = -2kx_L + kx_R, \tag{15.15a}$$

$$m\ddot{x}_R = kx_L - 2kx_R. \tag{15.15b}$$

Exercise 15.7b

Verify the results in Eqs. (15.15).

To find the normal modes for this system, we assume a solution of the form

$$x_L = a_L \cos(\omega t) \tag{15.16a}$$

$$x_R = a_R \cos(\omega t) \tag{15.16b}$$

for some amplitudes a_L, a_R, and angular frequency ω. Plugging the expressions (15.16) into the differential equations (15.15) and cancelling the common factors of $\cos(\omega t)$, we find

$$-m\omega^2 a_L = -2ka_L + ka_R, \tag{15.17a}$$

$$-m\omega^2 a_R = ka_L - 2ka_R. \tag{15.17b}$$

Divide through by $-k$ and write this in matrix notation as

$$\begin{pmatrix} 2 & -1 \\ -1 & 2 \end{pmatrix} \begin{pmatrix} a_L \\ a_R \end{pmatrix} = (m\omega^2/k) \begin{pmatrix} a_L \\ a_R \end{pmatrix}. \tag{15.18}$$

This is an eigenvalue–eigenvector problem, $Mv = \lambda v$, with matrix $M = \begin{pmatrix} 2 & -1 \\ -1 & 2 \end{pmatrix}$, eigenvalues $\lambda \equiv m\omega^2/k$, and eigenvectors $v = \begin{pmatrix} a_L \\ a_R \end{pmatrix}$.

The command `eval,evec = linalg.eig(M)` yields

```
array([3., 1.])
```

for the eigenvalues `eval` and

```
matrix([[ 0.7071068, 0.70710678],
        [-0.7071068, 0.70710678]])
```

for the eigenvectors `evec`. Since M is a 2×2 matrix, there are two eigenvalues and two eigenvectors. The first eigenvalue `eval[0]` and its corresponding eigenvector `evec[:,0]` are

$$\lambda_1 = 3.0, \tag{15.19a}$$

$$v_1 = \begin{pmatrix} 0.7071068 \\ -0.7071068 \end{pmatrix}. \tag{15.19b}$$

The second eigenvalue `eval[1]` and its corresponding eigenvector `evec[:,1]` are

$$\lambda_2 = 1.0, \tag{15.20a}$$

$$v_2 = \begin{pmatrix} 0.7071068 \\ 0.7071068 \end{pmatrix}. \tag{15.20b}$$

Note that `eval` is a NumPy array, whereas `evec` is a NumPy matrix.[2]

The two solutions of the eigenvalue–eigenvector problem are the *normal modes*. Let's take a close look at the first normal mode, defined by λ_1 and v_1. The amplitudes from Eqs. (15.16) are the components of v_1; thus, $a_L = 0.7071068$ and $a_R = -0.7071068$. The frequency (which we denote ω_1) is given by $\lambda_1 = m\omega_1^2/k$, so that $\omega_1 = \sqrt{k\lambda_1/m}$. To be concrete, let's assume the parameter values $m = 0.3\,\text{kg}$ and $k = 2\,\text{kg/s}^2$. Then the frequency for the first normal mode, with $\lambda_1 = 3.0$, is $\omega_1 = 4.47/\text{s}$. This corresponds to an oscillation period of $2\pi/\omega_1 = 1.40\,\text{s}$.

[2]It can be useful to define the eigenvectors as `v1 = evec[:,0]`, etc. This defines `v1` as a NumPy matrix with two rows and one column. In particular, `v1` is *not* a one-dimensional NumPy array. Since it takes two indices to access the elements of a matrix, the elements of `v1` are `v1[0,0]` and `v1[1,0]`.

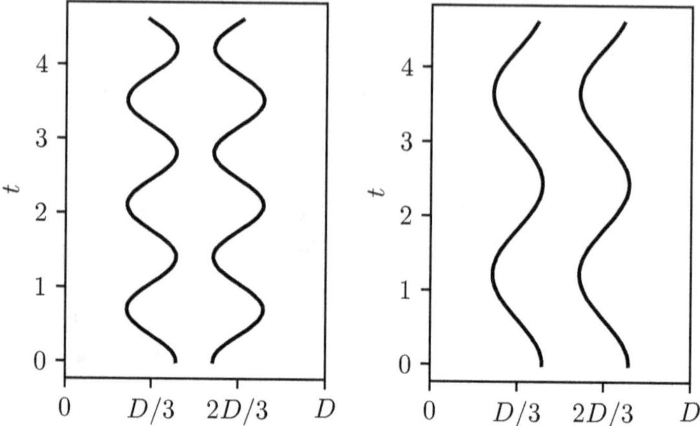

Fig. 15.6. The normal modes for the system of two masses and three springs shown in Fig. 15.5.

The second normal mode is defined by λ_2 and v_2. The amplitudes are $a_L = 0.7071068$ and $a_R = 0.7071068$ and the frequency is $\omega_2 = \sqrt{k\lambda_2/m} = 2.58/\mathrm{s}$. The oscillation period is 2.43 s.

The normal modes are shown in the two graphs of Fig. 15.6. For each mode, the positions X_L and X_R are plotted versus time t, with time on the vertical axis.

The general motion for this system is a linear combination of normal modes with arbitrary phase. That is, the general solution of the differential equations (15.14) is

$$\begin{pmatrix} X_L(t) \\ X_R(t) \end{pmatrix} = \begin{pmatrix} D/3 \\ 2D/3 \end{pmatrix} + A_1 v_1 \cos(\omega_1 t + \phi_1) + A_2 v_2 \cos(\omega_2 t + \phi_2), \tag{15.21}$$

where A_1 and A_2 are constant amplitudes and ϕ_1 and ϕ_2 are constant phase angles.

Exercise 15.7c

Verify the results of this section: Write a code to compute the normal modes for the system (15.14). Have your code reproduce the plots from Fig. 15.6.

Exercise 15.7d

Two masses (both m) are suspended from a ceiling by two springs, as shown in Fig. 15.7. The springs are identical, with stiffness k and relaxed length ℓ_0. Write a code to compute and plot the normal modes.

Exercise 15.7e

Three masses (each of mass m) move in one dimension along a frictionless table, as shown in Fig. 15.8. The masses are connected between fixed walls by four identical springs. Find the normal modes for this system. Plot the modes as functions of time.

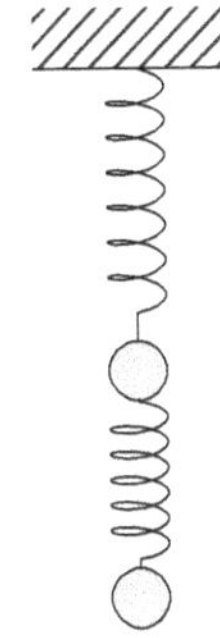

Fig. 15.7. Two masses suspended from the ceiling with springs.

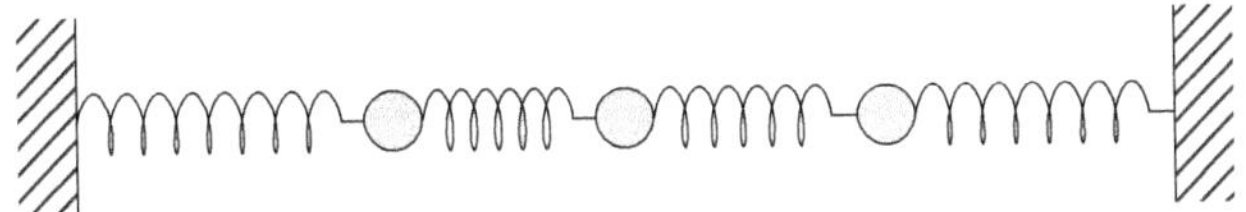

Fig. 15.8. Three masses connected between walls by four springs.

Chapter 16

Numerical Differentiation

Estimating the derivative of a function is a common task in scientific computing. The need arises when we have data that represent some quantity $f(x)$, dependent on a variable x, and we would like to know the rate of change $f'(x)$. If the data are obtained from an experiment or a numerical simulation, the function $f(x)$ is only known at discrete values of x. We must resort to numerical techniques to determine its derivative.

16.1 Two-point difference formulas

Our goal is to approximate the derivative of a function $f(x)$ at some point x_0. The only information we have is the value of $f(x)$ at discrete points along the x-axis. To be precise, $f(x)$ is known at x_0 and points separated from x_0 by multiples of h, as shown in Fig. 16.1. An approximation to $f'(x_0)$ or $f''(x_0)$ (or higher order derivatives) using the values $f(x_0)$, $f(x_0 \pm h)$, $f(x_0 \pm 2h)$, etc. is referred to as a *finite difference formula.*

Let's derive a simple finite difference formula for the first derivative $f'(x_0)$. Start with the Taylor series for $f(x)$ about the point x_0,

$$f(x) = f(x_0) + f'(x_0)(x - x_0) + \frac{1}{2}f''(x_0)(x - x_0)^2 + \cdots , \qquad (16.1)$$

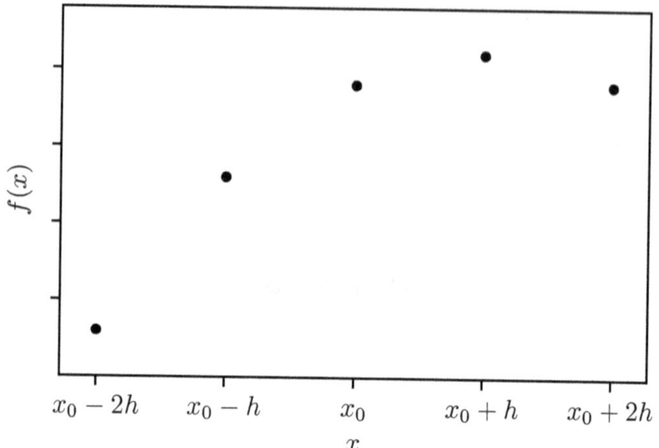

Fig. 16.1. Values of the function $f(x)$ are known at x_0, $x_0 \pm h$, $x_0 \pm 2h$, etc. We want to estimate $f'(x_0)$, the derivative of $f(x)$ at the point x_0.

where $\cdots$ represents terms proportional to $(x - x_0)^3$ and higher powers of $(x - x_0)$. Set $x = x_0 + h$ in this expression to obtain

$$f(x_0 + h) = f(x_0) + f'(x_0)h + \frac{1}{2}f''(x_0)h^2 + \cdots . \tag{16.2}$$

We can rearrange this relation to find

$$f'(x_0) = \frac{f(x_0 + h) - f(x_0)}{h} - \frac{1}{2}f''(x_0)h + \cdots . \tag{16.3}$$

Here, $\cdots$ represents terms proportional to h^2 and higher powers of h. Now, assuming h is small, the terms $-\frac{1}{2}f''(x_0)h + \cdots$ will be small. We can drop these terms and approximate the derivative of $f(x)$ at x_0 by

$$f'(x_0) \approx \frac{f(x_0 + h) - f(x_0)}{h}. \tag{16.4}$$

This formula for $f'(x_0)$ uses the values of $f(x)$ at just two points, x_0 and $x_0 + h$. We call this the *two-point forward difference* approximation to the derivative. Geometrically, the two-point forward difference is the slope of a straight line connecting the function values between x_0 and $x_0 + h$.

The error in the two-point forward difference comes from the terms $-\frac{1}{2}f''(x_0)h + \cdots$ that were dropped from Eq. (16.3). For small h, the unwritten terms $\cdots$ are small compared to $-\frac{1}{2}f''(x_0)h$. Thus, the error in the two-point forward difference formula is approximately proportional to h. This implies $|\text{error}| \approx ch$ for some constant c. Taking the logarithm of both sides, we have

$$\log|\text{error}| \approx \log c + \log h. \tag{16.5}$$

This result forms the basis of a convergence test for any code that uses the two-point forward difference formula: A plot of $\log|\text{error}|$ versus $\log h$ should approach a straight line with slope 1 as the resolution is increased.

The *two-point backwards difference* approximation is similar. Start with the Taylor expansion (16.1) and let $x = x_0 - h$ to obtain

$$f'(x_0) \approx \frac{f(x_0) - f(x_0 - h)}{h}. \tag{16.6}$$

This approximates the derivative by the slope of a line between x_0 and $x_0 - h$. The error in this approximation is proportional to h. Reducing h by a factor of 2 reduces the error by (approximately) a factor of 2.

Let's look at a simple example. The data that appear in Fig. 16.1 are: $f(0.8) = 0.8$, $f(2.2) = 1.8$, $f(3.6) = 2.4$, $f(5.0) = 2.6$, and $f(6.4) = 2.4$. The two-point forward and backward difference approximations to $f'(x)$ at $x = 3.6$ are

$$f'(3.6)\big|_{\text{forward}} \approx \frac{f(5.0) - f(3.6)}{5.0 - 3.6} = 0.143,$$

$$f'(3.6)\big|_{\text{backward}} \approx \frac{f(3.6) - f(2.2)}{3.6 - 2.2} = 0.429.$$

We can apply the difference formulas at other points in the data set as well. For example, the forward difference approximation at $x = 0.8$ is

$$f'(0.8)\big|_{\text{forward}} \approx \frac{f(2.2) - f(0.8)}{2.2 - 0.8} = 0.714,$$

and the backward difference approximation at $x = 6.4$ is

$$f'(6.4)\Big|_{\text{backward}} \approx \frac{f(6.4) - f(5.0)}{6.4 - 5.0} = -0.143.$$

Note that we cannot compute the forward difference approximation at $x = 6.4$, or the backward difference approximation at $x = 0.8$, because we don't have the necessary data.

Exercise 16.1

Consider the function $f(x) = \cos(x) + \sin(3x)$. Find the exact value of the derivative of $f(x)$ at the point $x_0 = 2$. Carry out a convergence test by computing the two-point forward difference approximation to $f'(2)$ with various values of h between 0.001 and 0.0001. Show that the errors are (approximately) proportional to h by plotting $\log|\text{error}|$ versus $\log h$.

16.2 Truncation errors and machine errors

The derivative of a function $f(x)$ at a point x_0 is defined by

$$f'(x_0) = \lim_{h \to 0} \frac{f(x_0 + h) - f(x_0)}{h}. \tag{16.7}$$

Apart from the limit symbol, this definition looks like the two-point forward difference formula (16.4). In fact, if we could take the limit as $h \to 0$ in the forward difference formula, the result would be the exact derivative, not just an approximation. Why can't we take the limit $h \to 0$, or at least take h to be very, very small?

In scientific computation, we usually don't have control over h. The value for h is dictated by the practical details of an experiment or a numerical simulation. Even if we could choose h freely, finite difference formulas are subject to machine roundoff error.

Let's look at an example with the function $f(x) = \sin x$, and consider its derivative at $x_0 = 1$. The exact answer is $f'(1) = \cos(1)$. Figure 16.2 shows a graph of $\log|\text{error}|$ versus $\log h$. As h is decreased from 10^{-1}, the error initially decreases. As h is decreased beyond about 10^{-8}, the error begins to grow.

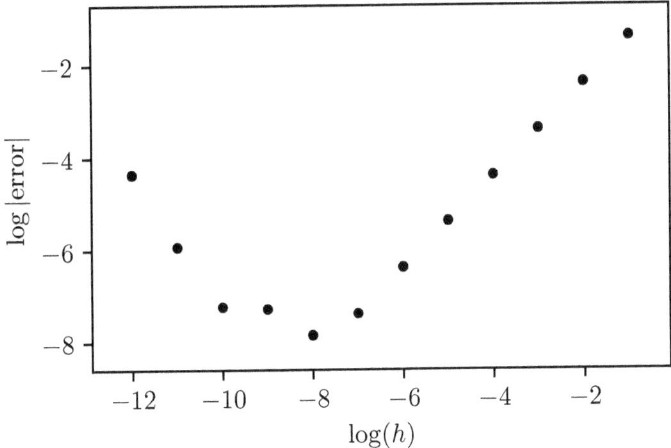

Fig. 16.2. Error as a function of h for the two-point forward difference approximation to $f'(x_0)$ with $f(x) = \sin x$ and $x_0 = 1$.

The analysis of the preceding section showed that the error in the two-point forward difference formula is $-\frac{1}{2}f''(x_0)h + \cdots$. We refer to this type of error as *truncation error* because it arises from truncating the infinite series (16.3) to obtain Eq. (16.4). With $f(x) = \sin x$, the truncation error at $x_0 = 1$ is approximately $\frac{1}{2}\sin(1)h$, or

$$\text{truncation error} \approx 0.4207\, h. \tag{16.8}$$

Now, the two-point forward difference formula is

$$f'(1) \approx \frac{\sin(1+h) - \sin(1)}{h}. \tag{16.9}$$

In addition to the truncation error, this calculation is subject to machine roundoff errors. As discussed in Sec. 4.4, a 64-bit machine can only represent the numerator $\sin(1+h) - \sin(1)$ to an accuracy of about $\pm 10^{-16}$. Thus, there are machine roundoff errors in computing the finite difference formula for $f'(1)$ of roughly $\pm 10^{-16}/h$. Table 16.1 compares the truncation error and machine roundoff error for various values of h. (Only the rough "order of magnitude" values are shown.) As h becomes smaller, truncation error is reduced but machine roundoff error grows. With a two-point forward difference approximation to $d\sin(x)/dx$ at $x = 1$, the best we can do is an overall error of about $\pm 10^{-8}$.

Table 16.1. Truncation error $\approx 0.4207\,h$ and roundoff error $\approx 10^{-16}/h$ for the two-point forward difference approximation to the derivative $f'(1)$, where $f(x) = \sin x$.

h	Truncation error	Roundoff error
10^{-6}	10^{-6}	10^{-10}
10^{-7}	10^{-7}	10^{-9}
10^{-8}	10^{-8}	10^{-8}
10^{-9}	10^{-9}	10^{-7}
10^{-10}	10^{-10}	10^{-6}
10^{-11}	10^{-11}	10^{-5}

Exercise 16.2

Use the two-point forward difference formula to compute the derivative of the function $f(x) = \cos(x) + \sin(3x)$ at $x_0 = 2$. (See Exercise 16.1.) Plot $\log|\text{error}|$ versus $\log h$ for $h = 10^{-2}$, 10^{-3}, etc. Do the errors always decrease as you decrease h?

16.3 Two-point forward difference, again

We can derive the two-point forward difference approximation (16.4) in a different way. Start with a linear combination of the relations

$$f(x_0) = f(x_0), \tag{16.10a}$$

$$f(x_0 + h) = f(x_0) + f'(x_0)h + \frac{1}{2}f''(x_0)h^2 + \cdots \tag{16.10b}$$

with coefficients a and b, which yields

$$af(x_0) + bf(x_0 + h) = (a + b)f(x_0) + bf'(x_0)h + \frac{b}{2}f''(x_0)h^2 + \cdots. \tag{16.11}$$

Now ask: What values should we choose for the coefficients a and b such that the right-hand side is as close as possible to $f'(x_0)$?

The answer is that the coefficient of $f(x_0)$ should vanish and the coefficient of $f'(x_0)$ should equal 1. That is, a and b should satisfy

$$a + b = 0, \tag{16.12a}$$

$$bh = 1. \tag{16.12b}$$

It would be nice if the coefficient of $f''(x_0)$ vanished, but that would give us a third equation $bh^2/2 = 0$ that is incompatible with Eq. (16.12b). The best we can do with just two unknowns, a and b, is to impose the two conditions (16.12).

The solution to Eqs. (16.12) is $b = 1/h$ and $a = -1/h$. With these values for the coefficients, Eq. (16.11) becomes

$$-\frac{1}{h} f(x_0) + \frac{1}{h} f(x_0 + h) = f'(x_0) + \frac{1}{2} f''(x_0)h + \cdots . \tag{16.13}$$

Assuming h is small, we can drop the terms $\frac{1}{2}f''(x_0)h + \cdots$ from the right-hand side. The result is the two-point forward difference formula, Eq. (16.4).

Exercise 16.3

Use the approach of this section to derive the two-point backward difference formula for $f'(x_0)$.

16.4 Centered difference formulas

We can use this same strategy to create other finite difference formulas. Start with the Taylor series expansions

$$f(x_0 - h) = f(x_0) - f'(x_0)h + \frac{1}{2} f''(x_0)h^2 + \cdots , \tag{16.14a}$$

$$f(x_0) = f(x_0), \tag{16.14b}$$

$$f(x_0 + h) = f(x_0) + f'(x_0)h + \frac{1}{2} f''(x_0)h^2 + \cdots , \tag{16.14c}$$

and form a linear combination with coefficients a, b, and c. The result is

$$\begin{aligned} &af(x_0-h)+bf(x_0)+cf(x_0+h) \\ &\quad = (a+b+c)f(x_0)+(c-a)f'(x_0)h+\frac{c+a}{2}f''(x_0)h^2+\cdots . \end{aligned} \tag{16.15}$$

Choose the coefficients such that the right-hand side is as close to $f'(x_0)$ as possible. We see that the coefficients of $f(x_0)$ and $f''(x_0)$ should vanish, and the coefficient of $f'(x_0)$ should be 1:

$$(a+b+c)=0, \tag{16.16a}$$

$$(c-a)h=1, \tag{16.16b}$$

$$(c+a)h^2/2=0. \tag{16.16c}$$

The solution is $a=-1/(2h)$, $b=0$ and $c=1/(2h)$, and Eq. (16.15) becomes

$$-\frac{1}{2h}f(x_0-h)+\frac{1}{2h}f(x_0+h)=f'(x_0)+\cdots . \tag{16.17}$$

If we assume h is small and drop the $\cdots$ terms, the result is

$$f'(x_0)\approx\frac{f(x_0+h)-f(x_0-h)}{2h}. \tag{16.18}$$

This is a *three-point centered difference* approximation for $f'(x_0)$. (The formula uses only two of the three points, because the coefficient of $f(x_0)$ vanishes.) This approximation to $f'(x_0)$ is simply the average of the two-point forward and backward difference formulas.

Observe that the unwritten terms in Eq. (16.15), the terms represented by $\cdots$, are proportional to h^3 multiplied by some combination of coefficients a, b and c. Since a, b, and c are themselves proportional to $1/h$, the unwritten terms are proportional to h^2. It follows that the truncation error in the centered difference formula (16.18) is proportional to h^2. That is, $|\text{error}|=Ch^2$ for some constant C and a plot of $\log|\text{error}|$ versus $\log h$ should be (approximately) a straight line with slope 2.

Geometrically, the three-point centered difference formula (16.18) uses the slope of the line connecting the data points at x_0-h and x_0+h to approximate $f'(x_0)$.

Exercise 16.4a

Consider the function $f(x) = \cos(x)\exp(-x^2/2)$. Use the three-point centered difference with $h = 0.1$ to approximate $f'(x)$ at discrete points in the domain $-3 \leq x \leq 3$. Plot your approximate results for $f'(x)$ and, on the same graph, the exact function $f'(x)$.

Return to the linear combination in Eq. (16.15). We can use this to derive a finite difference formula for the second derivative, $f''(x_0)$, by setting the coefficients of $f(x_0)$ and $f'(x_0)$ equal to zero and the coefficient of $f''(x_0)$ equal to 1. That is,

$$(a + b + c) = 0, \tag{16.19a}$$

$$(c - a)h = 0, \tag{16.19b}$$

$$(c + a)h^2/2 = 1, \tag{16.19c}$$

which has the solution $a = 1/h^2$, $b = -2/h^2$ and $c = 1/h^2$. This gives

$$\frac{1}{h^2} f(x_0 - h) - \frac{2}{h^2} f(x_0) + \frac{1}{h^2} f(x_0 + h) = f''(x_0) + \cdots . \tag{16.20}$$

After dropping the $\cdots$ terms, we have the three-point centered difference approximation to the second derivative of $f(x)$:

$$f''(x_0) \approx \frac{f(x_0 - h) - 2f(x_0) + f(x_0 + h)}{h^2}. \tag{16.21}$$

If we keep track of the unwritten terms in this derivation, we find that the truncation error is proportional to h^2.

Exercise 16.4b

Use the three-point centered difference approximation to compute $f''(x_0)$, where $f(x) = \sin x$ and $x_0 = 1$. Carry out a convergence test with $0.1 \leq h \leq 0.0001$ to show that the error is (approximately) proportional to h^2. What happens to the error as h decreases below 0.0001?

Exercise 16.4c

The data file *ToDifferentiateData.txt* (see the Appendix) contains two columns, x and $f(x)$, with x values equally spaced. Use the three-point centered difference formulas to approximate the first and second derivatives of $f(x)$ at each of the data points, excluding the endpoints. Graph your results.

16.5 Other stencils

The pattern of evaluation points for a finite difference formula is sometimes referred to as a *stencil.* For the two-point forward difference formula, the stencil consists of the points x_0 and $x_0 + h$. For the three-point centered difference formulas, the stencil consists of $x_0 - h$, x_0 and $x_0 + h$.

We can use various stencils to derive more finite difference formulas for approximating derivatives. For example, with the three-point stencil x_0, $x_0 + h$, $x_0 + 2h$, we have

$$f(x_0) = f(x_0), \tag{16.22a}$$

$$f(x_0 + h) = f(x_0) + f'(x_0)h + \frac{1}{2}f''(x_0)h^2 + \cdots, \tag{16.22b}$$

$$f(x_0 + 2h) = f(x_0) + 2f'(x_0)h + 2f''(x_0)h^2 + \cdots. \tag{16.22c}$$

A linear combination yields

$$\begin{aligned} &af(x_0) + bf(x_0 + h) + cf(x_0 + 2h) \\ &\quad = (a + b + c)f(x_0) + (b + 2c)f'(x_0)h + (b/2 + 2c)f''(x_0)h^2 + \cdots. \end{aligned} \tag{16.23}$$

To approximate the first derivative, $f'(x_0)$, choose

$$(a + b + c) = 0, \tag{16.24a}$$

$$(b + 2c)h = 1, \tag{16.24b}$$

$$(b/2 + 2c)h^2 = 0, \tag{16.24c}$$

with the solution $a = -3/(2h)$, $b = 2/h$, $c = -1/(2h)$. This leads to the three-point, one sided approximation

$$f'(x_0) \approx \frac{-3f(x_0) + 4f(x_0 + h) - f(x_0 + 2h)}{2h}. \tag{16.25}$$

In this case, the truncation error is proportional to h^2.

We can also approximate the second derivative $f''(x_0)$ using the stencil x_0, $x_0 + h$, $x_0 + 2h$. Choose

$$(a + b + c) = 0, \tag{16.26a}$$

$$(b + 2c)h = 0, \tag{16.26b}$$

$$(b/2 + 2c)h^2 = 1. \tag{16.26c}$$

which has the solution $a = 1/h^2$, $b = -2/h^2$, $c = 1/h^2$. The resulting approximation is

$$f''(x_0) \approx \frac{f(x_0) - 2f(x_0 + h) + f(x_0 + 2h)}{h^2}, \tag{16.27}$$

with truncation error proportional to h.

Exercise 16.5a

Show that the finite difference approximations for first and second derivatives with stencil $x_0 - 2h$, $x_0 - h$, x_0 are

$$f'(x_0) \approx \frac{f(x_0 - 2h) - 4f(x_0 - h) + 3f(x_0)}{2h},$$

$$f''(x_0) \approx \frac{f(x_0 - 2h) - 2f(x_0 - h) + f(x_0)}{h^2}.$$

These are three-point one sided approximations for $f'(x_0)$ and $f''(x_0)$, with stencil points in the backward direction. The approximations Eqs. (16.25) and (16.27) are one sided with stencil points in the forward direction.

Exercise 16.5b

Sample the function $f(x) = \sin(x + 2\sin(x))$ at N points in the domain $0 \le x \le 2\pi$, then use the data to compute the first derivative $f'(x)$ at each of the sample points. Use the three-point centered stencil for the interior points (the points between 0 and 2π), and use the three-point one sided stencils at the endpoints 0 and 2π. (See Eq. (16.25) and Exercise 16.5a.) Plot your finite difference approximation along with the actual expression for $f'(x)$ on a single graph. Experiment with different values of N.

Exercise 16.5c

Show that the five-point centered difference formula for the first derivative is

$$f'(x_0) \approx \frac{-f(x_0+2h) + 8f(x_0+h) - 8f(x_0-h) + f(x_0-2h)}{12h}.$$

(Suggestion: Use SymPy to solve the system of five equations for the five coefficients.)

Exercise 16.5d

Derive the five-point centered difference formula for the second derivative, $f''(x_0)$.

Chapter 17

Ordinary Differential Equations I

17.1 Newton's law of cooling

Newton's law of cooling says that the rate of change of temperature of a body is proportional to the temperature difference between the body and its surroundings. Let $T(t)$ denote the temperature of the body as a function of time t. Newton's law of cooling is

$$\frac{dT(t)}{dt} = -k[T(t) - T_a], \tag{17.1}$$

where T_a is the ambient temperature, the temperature of the surroundings. The left-hand side of this equation is the time rate of change of the body's temperature. The right-hand side is the temperature difference, $T(t) - T_a$, multiplied by a proportionality constant $-k$ with $k > 0$. If $T(t) > T_a$, then $dT(t)/dt$ is negative and $T(t)$ decreases in time. If $T(t) < T_a$, then $dT(t)/dt$ is positive and $T(t)$ increases in time. The particular value for k depends on the size, shape and composition of the body, as well as the nature of the surroundings.

Equation (17.1) is an ordinary differential equation (ODE) for the unknown function $T(t)$. Most differential equations cannot be solved analytically, but Newton's law of cooling is an exception. The exact, analytical solution is found by separation of variables:

$$\frac{dT}{T - T_a} = -k\,dt. \tag{17.2}$$

Integrating both sides, we have

$$\ln(T - T_a) = -k\,t + c, \tag{17.3}$$

where c is an integration constant. Solving for T gives

$$T(t) = T_a + e^c e^{-kt}. \tag{17.4}$$

Let $T_0 = T(0)$ denote the temperature of the body at the initial time $t = 0$. Then Eq. (17.4) evaluated at $t = 0$ implies $e^c = T_0 - T_a$. This yields

$$T(t) = T_a + (T_0 - T_a)e^{-kt} \tag{17.5}$$

for the temperature of the body as a function of time.

The "half-life" t_H for the cooling (or heating) process is the time it takes the body to reach a temperature that is half-way between the initial and ambient temperatures. That is, when $t = t_H$, the temperature is $T = (T_0 + T_a)/2$. Inserting these values into the solution (17.5), we find

$$k = \frac{1}{t_H}\ln 2, \tag{17.6}$$

where ln is the natural logarithm.

Exercise 17.1

Plot a graph of $T(t)$ versus t using reasonable values for the initial and ambient temperatures T_0 and T_a, and the half-life t_H.

17.2 Euler's method

Let's solve the differential equation (17.1) numerically. Our goal will be to find the temperature of the body at discrete times t_0, t_1, t_2, etc. up to some final time t_N. Let the index i range from 0 through N, so the discrete times are t_i. The time difference from one discrete time to the next is $\Delta t = t_{i+1} - t_i$, or $\Delta t = (t_N - t_0)/N$. See Fig. 17.1.

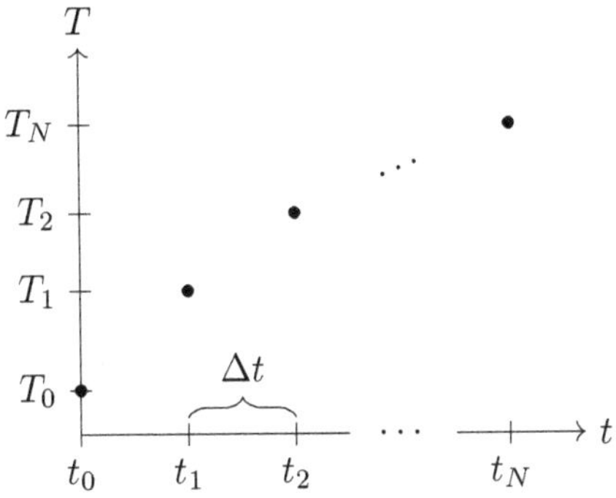

Fig. 17.1. Values of temperature $T_i = T(t_i)$ at discrete times t_i. The figure shows N timesteps, each of length Δt.

At each of the times t_i the rate of change of temperature should equal $-k[T(t_i) - T_a]$. Recall from Ch. 16 on numerical differentiation that the derivative at time t_i can be approximated by

$$\dot{T}(t_i) \approx \frac{T(t_i + \Delta t) - T(t_i)}{\Delta t}, \tag{17.7}$$

where the "dot" denotes d/dt. This is the two point forward difference formula (16.4) with some changes of notation: T in place of f, t_i in place of x_0, Δt in place of h and "dot" in place of "prime". We can use this approximation to replace the derivative in Newton's law of cooling, with the result

$$\frac{T(t_i + \Delta t) - T(t_i)}{\Delta t} = -k[T(t_i) - T_a]. \tag{17.8}$$

Now simplify the notation by letting $T_i = T(t_i)$. Since $t_{i+1} = t_i + \Delta t$, then $T_{i+1} = T(t_{i+1}) = T(t_i + \Delta t)$ and Eq. (17.8) becomes

$$T_{i+1} = T_i - k(T_i - T_a)\Delta t. \tag{17.9}$$

The temperature at each discrete time t_i is obtained by iterating this equation. Each iteration is called a *timestep*. Given an initial temperature T_0, we use Eq. (17.9) with $i = 0$ to obtain T_1. This is the first timestep. Now that we know T_1, we use Eq. (17.9) with $i = 1$ to obtain T_2. This is the second timestep. We continue in this fashion for N timesteps.

This step-by-step procedure for solving the differential equation (17.1) is called *Euler's method.* Here is an example code using $T_a = 180°\text{C}$, $T_0 = 20°\text{C}$, $t_H = 1000\,\text{s}$, and a final time of 3600 s:

```
import numpy as np

tH = 1000.0            # half--life in seconds (s)
Ta = 180.0             # ambient temperature in Celcius (C)
Tinitial = 20.0        # initial temperature (C)
tinitial = 0.0         # initial time (s)
tfinal = 3600.0        # final time (s)
N = 20                 # number of timesteps

Deltat = (tfinal - tinitial)/N  # compute Delta t
k = np.log(2)/tH                # compute k

# Create arrays to hold times and temperatures
t = np.linspace(tinitial,tfinal,N+1)
T = np.zeros(len(t))

# Solve using Euler's method
T[0] = Tinitial            # initial temperature
for i in range(N):         # loop for N timesteps
    T[i+1] = T[i] - k*(T[i] - Ta)*Deltat

print(T)                   # print temperature array
```

Make sure you understand how this code works. Why does `N+1` appear in the `linspace` command if there are only `N` timesteps?

Exercise 17.2

Use Euler's method to solve the differential equation (17.1) for Newton's law of cooling. Use the parameters listed above. Experiment with different values of N and compare the numerical answers for the final temperature to the analytical result $T(3600) = 166.80°\text{C}$. The numerical answers should improve as the number of timesteps is increased. Make a plot of temperature versus time showing both the numerical data and the analytical result for $T(t)$.

17.3 Truncation error

Euler's method is subject to truncation errors because it uses Eq. (17.9) iteratively to approximate the solution of the differential equation (17.1). The truncation errors come from the use of the two-point forward difference formula (17.7) in place of the time derivative $\dot{T}$. The two-point forward difference formula is a truncated version of the exact relation

$$\dot{T}(t_i) = \frac{T_{i+1} - T_i}{\Delta t} - \frac{1}{2}\ddot{T}(t_i)\Delta t + \cdots . \tag{17.10}$$

This is just Eq. (16.3) with, once again, some changes in notation. It is obtained by expanding $T(t_i + \Delta t)$ in a Taylor series about $T(t_i)$.

Let's rederive the iterative formula (17.9), this time keeping the extra terms from Eq. (17.10). Replace the time derivative in Newton's law of cooling with the right-hand side of Eq. (17.10), then rearrange:

$$T_{i+1} = T_i - k(T_i - T_a)\Delta t + \frac{1}{2}\ddot{T}(t_i)\Delta t^2 + \cdots . \tag{17.11}$$

The $\cdots$ terms are proportional to Δt^3 and higher powers of Δt.

This result shows that the error for a single timestep of the Euler method is

$$\text{error for one timestep} = \frac{1}{2}\ddot{T}(t_i)\Delta t^2 + \cdots . \tag{17.12}$$

Assuming Δt is small we can drop the $\cdots$ terms. Then the total error for N timesteps is

$$\text{Euler's method error} \approx \sum_{i=0}^{N-1} \frac{1}{2}\ddot{T}(t_i)\Delta t^2. \tag{17.13}$$

Apart from a factor of $\Delta t/2$, we can recognize this expression as the left-endpoint rule approximation to the integral of $\ddot{T}$. That is,

$$\text{Euler's method error} \approx \frac{\Delta t}{2}\int_{t_0}^{t_N} \ddot{T}\, dt. \tag{17.14}$$

The integral equals $\dot{T}(t_N) - \dot{T}(t_0)$. This fact is not particularly useful because (in general) we don't know the values of $\dot{T}(t_N)$ or $\dot{T}(t_0)$.

Nevertheless, these results show that the error for Euler's method is proportional to Δt. In turn, Δt is proportional to N^{-1}, where N is the number of timesteps. Therefore, the error in the final value of T is

$$\text{Euler's method error} \propto N^{-1}. \tag{17.15}$$

You can perform a convergence test of your Euler's method code by finding the error in the final value of temperature as a function of resolution N.

Exercise 17.3a

Use Euler's method to solve Newton's law of cooling (17.1) with the parameter values of Sec. 17.2. Compute the error in the temperature at the final time $t = 3600\,\text{s}$ for various values of N, and plot $\log|\text{error}|$ versus $\log(N)$. Find the slope from the plot and verify that the error is proportional to N^{-1}.

Clearly, there is nothing special about 3600 s. We could choose a final time of, say, 1800 s and the conclusion would be the same: The error at any time is proportional to N^{-1}.

We can analyze the error at 1800 s even if the final time used in the simulation is 3600 s.

Exercise 17.3b

Repeat the analysis of the previous exercise. Keep the final run time at 3600 s, but compute the error at 1800 s. Plot $\log|\text{error}|$ versus $\log(N)$. Should you use the total number of timesteps N needed to reach 3600 s, or half that number? Show that it doesn't matter. Either way, you should find that the error is proportional to N^{-1}.

17.4 Error estimation

Previously, we assumed that the ambient temperature T_a in Newton's law of cooling is a constant. We can modify the problem by letting the ambient temperature vary in time. For example, let

$$T_a(t) = a + b\arctan(t/c), \tag{17.16}$$

where a, b, and c are constants and arctan is the inverse tangent function. Newton's law of cooling becomes

$$\dot{T}(t) = -k[T(t) - a - b\arctan(t/c)]. \tag{17.17}$$

This differential equation can't be solved analytically, but we can still solve it numerically. Using Euler's method the differential equation becomes

$$T_{i+1} = T_i - k[T_i - a - b\arctan(t_i/c)]\Delta t. \tag{17.18}$$

To be concrete, let the constants have values $a = 20°\text{C}$, $b = 100°\text{C}$, and $c = 1000\,\text{s}$. Also choose $T_0 = 60°\text{C}$ for the initial temperature and $k = 0.002/\text{s}$. What is the body's temperature at time $t = 3000\,\text{s}$?

Figure 17.2 shows a plot of the temperatures T_i obtained from Euler's method with $N = 10$. The plot also shows the ambient temperature $T_a(t)$ as a function of time.

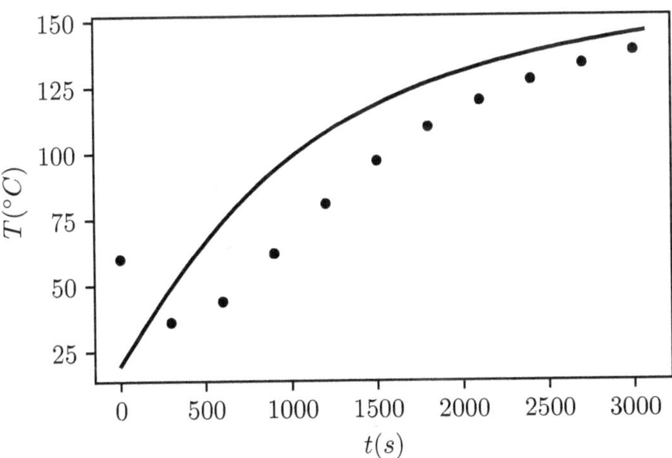

Fig. 17.2. Newton's law of cooling for a body with initial temperature $T_0 = 60°\text{C}$ and ambient temperature $T_a(t) = 20°\text{C} + 100°\text{C}\arctan(t/1000\,\text{s})$. The dots are obtained from Euler's method with $N = 10$ timesteps. The solid curve is the ambient temperature.

Let's take a close look at the temperature at the final time, $t_N = 3000\,\text{s}$. With Euler's method and $N = 10$ timesteps, we find that the final temperature is $T(3000) = 138.384084°\text{C}$. How accurate is this result? If we only run our code with $N = 10$ timesteps, there is no way to tell how accurate the answer is. Since the truncation errors are proportional to N^{-1}, we can improve the result by increasing N. If we run the code with $N = 20$, the numerical answer changes from 138.384084° C to 137.850497° C. This tells us that the correct answer is around 137°C or 138°C. To be sure, we should run the code at even higher resolution.

Table 17.1 shows the final temperatures for increasing values of N. By $N = 80$, we see that the correct answer to three significant figures is 137° C and to four significant figures the answer is approximately 137.4° C. To be confident in the answer to four significant figures, we need to increase the resolution even further. By $N = 1280$, it is apparent that the final temperature is 137.3° C. By $N = 2560$, the answer has settled down to 137.34° C, giving us five significant figures.

Remember: You can't tell how accurate your numerical answers are until you run your code at multiple resolutions.

Table 17.1. Temperatures at the final time $t_N = 3000\,\text{s}$ for a body with initial temperature $T_0 = 60°\text{C}$, in an environment with ambient temperature given by Eq. (17.16).

N	final temp (°C)
10	138.384084
20	137.850497
40	137.588677
80	137.461121
160	137.398355
320	137.367243
640	137.351756
1280	137.344030
2560	137.340172

Exercise 17.4a

Use Euler's method to solve for the temperature of a body when the ambient temperature varies sinusoidally:

$$T_a(t) = a + b\sin(t/c).$$

Choose $a = 100°\mathrm{C}$, $b = 40°\mathrm{C}$, and $c = 1000\,\mathrm{s}$. Also let $T_0 = 140°\mathrm{C}$ and $k = 0.001$. Find the temperature of the body at time 16000 s, accurate to 4 significant figures.

Exercise 17.4b

As in the previous exercise, let the ambient temperature vary sinusoidally with $a = 100°\mathrm{C}$, $b = 40°\mathrm{C}$, $c = 1000\,\mathrm{s}$, and $T_0 = 140°\mathrm{C}$. Experiment with different values of k. Plot a graph showing both T and T_a as functions of time. As time passes, the body temperature T should settle down and execute regular oscillations. How does k affect the oscillation amplitude for T? How does k affect the phase difference between T and T_a? Interpret these results physically.

17.5 Three-point convergence test

Continue to focus on the final temperatures listed in Table 17.1. Let $T_{(N)}$ denote the temperature at the final time obtained from Euler's method with N timesteps. That is, $T_{(10)} = 138.384084°\mathrm{C}$, $T_{(20)} = 137.850497°\mathrm{C}$, etc. The differences between successive temperature values are

$$T_{(10)} - T_{(20)} = 0.533587°\mathrm{C},$$

$$T_{(20)} - T_{(40)} = 0.261820°\mathrm{C},$$

$$T_{(40)} - T_{(80)} = 0.127556°\mathrm{C},$$

$$T_{(80)} - T_{(160)} = 0.062766°\mathrm{C},$$

$$T_{(160)} - T_{(320)} = 0.031112°\mathrm{C},$$

$$T_{(320)} - T_{(640)} = 0.015487°\text{C},$$
$$T_{(640)} - T_{(1280)} = 0.007726°\text{C},$$
$$T_{(1280)} - T_{(2560)} = 0.003858°\text{C}.$$

Do you see the pattern? These numbers differ by a factor of about 2 from one level to the next. For example, for the first two levels,

$$\frac{T_{(10)} - T_{(20)}}{T_{(20)} - T_{(40)}} = 2.03799, \tag{17.19}$$

and for the last two levels,

$$\frac{T_{(640)} - T_{(1280)}}{T_{(1280)} - T_{(2560)}} = 2.003. \tag{17.20}$$

This is not a coincidence, rather, a consequence of the fact that the errors are approximately proportional to N^{-1}.

Let's show why this should be the case. Express the final temperature as

$$T_{(N)} = T_{(\text{exact})} + C/N, \tag{17.21}$$

where $T_{(\text{exact})}$ is the (unknown) exact answer and C is a proportionality constant. This relation holds, approximately, for any number of timesteps. In particular, for $2N$ and $4N$ timesteps, we have

$$T_{(2N)} = T_{(\text{exact})} + C/(2N), \tag{17.22a}$$
$$T_{(4N)} = T_{(\text{exact})} + C/(4N). \tag{17.22b}$$

These last three equations combine to yield

$$\frac{T_{(N)} - T_{(2N)}}{T_{(2N)} - T_{(4N)}} = 2. \tag{17.23}$$

This suggests a simple test for any Euler's method code, called a *three-point convergence test*. Compute the expression (17.23) using three successive values for the number of timesteps. The result should be close to 2 in the limit as the three resolutions are increased. If it isn't, there is likely an error in the code.[1]

[1] For some systems, such as the driven damped pendulum of Ch. 19, a convergence test might fail due to the growth of machine roundoff errors.

Exercise 17.5

Apply the three-point convergence test to your code from Exercise 17.4a.

17.6 Richardson extrapolation

Since we know how the errors change as N increases, we can use this information to predict the temperature for values of N beyond 2560. Of course, what we really want is the value of T as $N \to \infty$. In this limit $T_{(N)}$ equals the exact value, $T_{(\mathrm{exact})}$. To find $T_{(\mathrm{exact})}$, we can use the relations

$$T_{(\mathrm{low})} = T_{(\mathrm{exact})} + C/(N_{\mathrm{low}}), \tag{17.24a}$$

$$T_{(\mathrm{high})} = T_{(\mathrm{exact})} + C/(N_{\mathrm{high}}), \tag{17.24b}$$

for two values of N, which we call N_{low} and N_{high}. Eliminating C and solving for the exact temperature gives

$$T_{(\mathrm{exact})} = 2T_{(\mathrm{high})} - T_{(\mathrm{low})}, \tag{17.25}$$

assuming $N_{\mathrm{high}} = 2N_{\mathrm{low}}$.

This result is known as *Richardson extrapolation.* In practice, we usually apply Richardson extrapolation to the two highest resolutions available. For our Newton's cooling problem with ambient temperature (17.16),

$$T_{(\mathrm{exact})} = 2T_{(2560)} - T_{(1280)} = 137.336313^\circ\mathrm{C}. \tag{17.26}$$

Of course, this isn't really the *exact* answer; keep in mind that Eqs. (17.24) are approximations that ignore terms proportional to higher powers of $1/N$. Nevertheless, the result from Richardson extrapolation should be more accurate than any of the individual results in Table 17.1.

Just how accurate is the result (17.26)? As usual, it's not possible to judge the accuracy without further information. Let's do a numerical experiment. Consider a code that uses Euler's method to solve

Newton's cooling with ambient temperature (17.16) using two values for the timestep, N_{low} and N_{high}. The code then uses Richardson extrapolation to find the final temperature. A few results are shown as follows:

N_{low}	N_{high}	$T(°\text{C})$
320	640	137.336269
640	1280	137.336304
1280	2560	137.336313

Evidently the final temperature is $T = 137.3363°\text{C}$ to 7 significant figures. To 8 significant figures the answer is probably 137.33631°C or 137.33632°C, but we need further tests to be sure.

Richardson extrapolation can be applied in many contexts, including numerical integration and differentiation. In general, our numerical schemes yield results with errors proportional to some power of $1/N$, where N is a measure of the resolution. For ODEs, N is the number of timesteps; for numerical integration, N is the number of subintervals; for numerical differentiation, $N \propto 1/h$, where h is the separation between stencil points. For a numerical method of "order n," the leading term in the error is proportional to $1/N^n$. In this case, we can compute the numerical approximations of some quantity Q using resolutions $N_{(\text{low})}$ and $N_{(\text{high})}$. The low and high Q values will be related to the "exact" answer by

$$Q_{(\text{low})} = Q_{(\text{exact})} + C/(N_{\text{low}})^n, \tag{17.27a}$$

$$Q_{(\text{high})} = Q_{(\text{exact})} + C/(N_{\text{high}})^n. \tag{17.27b}$$

Assuming $N_{\text{high}} = 2N_{\text{low}}$, these equations imply

$$Q_{(\text{exact})} = \frac{2^n Q_{(\text{high})} - Q_{(\text{low})}}{2^n - 1}. \tag{17.28}$$

For example, Simpson's rule is an order $n = 4$ scheme for numerical integration, with errors proportional to N^{-4}. You can use Simpson's rule to approximate an integral at resolutions N_{low} and

$N_{\text{high}} = 2N_{\text{low}}$, then use Richardson extrapolation (17.28) with $n = 4$ to obtain an improved answer.

Exercise 17.6a

Consider once again Newton's law of cooling with the sinusoidally varying ambient temperature of Eq. (17.16). Use the parameter values $a = 100°\text{C}$, $b = 30°\text{C}$, $c = 400\,\text{s}$, $T_0 = 10°\text{C}$, and $k = 0.005$. Solve this problem using Euler's method at two resolutions, N_{low} and $N_{\text{high}} = 2N_{\text{low}}$, and apply Richardson extrapolation. Determine the body temperature at time 8000 s with an accuracy of 6 significant figures.

Exercise 17.6b

Use Simpson's rule and Richardson extrapolation to approximate the integral

$$I = \int_0^{10} \ln(x + \cos(x))\, dx$$

to 8 significant figures.

Exercise 17.6c

Consider the function

$$f(x) = e^{\sin(x+x^2)}.$$

What is $f'(0)$? Use the centered 3-point stencil to approximate $f'(0)$ at high resolution $h = 0.1$, and at low resolution $h = 0.2$. Verify that the centered 3-point difference formula for $f'(x)$ is an order $n = 2$ approximation by showing that the error decreases by a factor of about 4 between the low and high resolution results. Use Richardson extrapolation to find an improved answer for $f'(0)$.

17.7 Practical issues with error estimation

We worry a lot about errors in numerical calculations. You might ask: Can I just use a really high resolution (large N) and simply assume the answer is "accurate enough?" This strategy might be fine if you want a rough answer to a simple problem. But remember, there are always machine roundoff errors. If your resolution is too high, roundoff errors can spoil the results. There are also practical limitations on resolution. Realistic research problems can be very complicated, and can stretch the limits of human patience and computer memory. You might not be able to reach a really high resolution because it simply takes too long for the calculation to finish, or because it overloads the available memory on your computer.

You might also reason: "I want an answer that is accurate to, say, 3 significant figures. I will instruct Python to print the answer to 3 significant figures, then increase the resolution until the numbers stop changing." Is this a valid strategy? The short answer is maybe. This strategy sometimes works, but not always. Consider an example problem with exact answer $Q_{(\text{exact})} = 1.4443$. If the numerical scheme has errors of order N^{-1}, then the numerical answers at resolution N have the form

$$Q_{(N)} = 1.4443 + C/N \tag{17.29}$$

for some constant C. Let's assume $C = 0.0321$. The results for Q with increasing N are shown in this table:

N	1	2	4	8	16	32	64	128
Q	1.48	1.46	1.45	1.45	1.45	1.45	1.44	1.44

If we were to follow the proposed strategy, we would see that the answer is unchanged from $N = 4$ to $N = 8$ and declare the answer to be 1.45. This is wrong; the correct answer to 3 significant figures, 1.44, does not appear until the resolution reaches $N = 64$. This strategy can be especially misleading if N is increased by a small amount from one numerical test to the next.

Exercise 17.7

Use Simpson's rule to approximate the integral

$$I = \int_0^{\pi} \sin(x)\, dx,$$

using $N = 10^9$ timesteps. How long does it take for your computer to complete the calculation? (Use the `time()` function.) The exact answer is $I = 2$. How does the error at $N = 10^9$ compare to the error at $N = 10^3$?

Chapter 18

Ordinary Differential Equations II

18.1 Systems of equations

Euler's method can be applied to any system of ordinary differential equations (ODEs). Consider the following example:

$$dx/dt = xy - z + t, \tag{18.1a}$$

$$dy/dt = 2z + x, \tag{18.1b}$$

$$dz/dt = t^2 - yz. \tag{18.1c}$$

This set of equations has one independent variable (namely t) and three dependent variables (namely, x, y, and z). The goal is to determine the functions $x(t)$, $y(t)$, and $z(t)$.

We can apply the same reasoning as in the previous chapter and replace the differential equations (18.1) with a set of discrete equations. This process is called *discretization.*

Begin by evaluating the differential equations at time t_i, then replace the derivatives such as dx/dt with the two-point forward difference approximation $(x_{i+1} - x_i)/\Delta t$. Rearrange the results to obtain

$$x_{i+1} = x_i + (x_i y_i - z_i + t_i)\Delta t, \tag{18.2a}$$

$$y_{i+1} = y_i + (2z_i + x_i)\Delta t, \tag{18.2b}$$

$$z_{i+1} = z_i + (t_i^2 - y_i z_i)\Delta t. \tag{18.2c}$$

These equations are solved iteratively. Step 1: Insert the initial values x_0, y_0, z_0 into the right-hand sides of Eqs. (18.2) with $i = 0$.

This yields x_1, y_1, z_1. Step 2: Insert x_1, y_1, z_1 into the right-hand sides of Eqs. (18.2) with $i = 1$. This yields x_2, y_2, z_2. Continue for N timesteps to reach x_N, y_N, z_N.

Exercise 18.1a

Solve the system (18.1) numerically using Euler's method (18.2) with initial data $x(0) = 0$, $y(0) = 0$, and $z(0) = 1$. Make a plot of the dependent variables (x, y, and z) versus t for $0 \le t \le 2.0$.

Exercise 18.1b

Use Euler's method to solve the system of ODEs

$$dx/dt = yt/x,$$
$$dy/dt = x - y + t,$$

with initial conditions $x(0) = -20$ and $y(0) = 5$. Plot x and y versus t on a single graph with $0 \le t \le 25$. Experiment with different numbers of timesteps N and determine the final values $x(25)$ and $y(25)$ to 5 significant figures.

18.2 Euler's method in general

Any system of first-order ODEs can be described using the shorthand notation

$$\frac{du}{dt} = F(u,t). \tag{18.3}$$

Here, u denotes the dependent variables (the unknown functions) and $F(u,t)$ denotes the right-hand sides. For example, consider the system (18.1) from the previous section. The dependent variables x, y, and z are collected into a "vector" of unknowns,

$$u(t) = \begin{pmatrix} x(t) \\ y(t) \\ z(t) \end{pmatrix}, \tag{18.4}$$

and the right-hand sides are written as

$$F(u,t) = \begin{pmatrix} xy - z + t \\ 2z + x \\ t^2 - yz \end{pmatrix} . \qquad (18.5)$$

With these identifications for $u(t)$ and $F(u,t)$, Eq. (18.3) is equivalent to Eqs. (18.1).

Euler's method is obtained by replacing the derivatives du/dt with the two–point forward difference formula $(u_{i+1} - u_i)/\Delta t$ and evaluating $F(u,t)$ at time t_i. Solving for u_{i+1}, we find

$$u_{i+1} = u_i + F(u_i, t_i)\Delta t. \qquad (18.6)$$

Here, $u_i = u(t_i)$ denotes the dependent variables at the discrete times t_i. Also recall that $\Delta t = (t_N - t_0)/N$, where N is the number of timesteps.

Exercise 18.2

Use Euler's method to solve the system of ODEs $du/dt = F(u,t)$ with

$$u = \begin{pmatrix} x \\ y \end{pmatrix}, \quad F(u,t) = \begin{pmatrix} y + t \\ -4(x - t) \end{pmatrix}.$$

Choose initial conditions $x(0) = 0$, $y(0) = 12$ and time interval $0 \le t \le 10$. Apply the three-point convergence test of Eq. (17.23) to each of the variables x and y.

18.3 Second-order equations

Up to this point we have focused on first-order ODEs. Higher order ODEs can be treated using the same numerical techniques by defining new variables. For example, consider the second-order ODE

$$\ddot{x} = 2xt - x\dot{x}, \qquad (18.7)$$

with initial conditions $x(0) = 1$ and $\dot{x}(0) = 0$. (Dots denote time derivatives.) We can write this as a system of first-order ODEs by defining a new variable $y = \dot{x}$. Equation (18.7) is equivalent to

$$\dot{x} = y, \tag{18.8a}$$

$$\dot{y} = 2xt - xy, \tag{18.8b}$$

with initial conditions $x(0) = 1$ and $y(0) = 0$.

Exercise 18.3a

Solve the system (18.8) with the given initial conditions using Euler's method. Plot a graph of x versus t for the domain $0 \le t \le 3$, and find the value of $x(3)$ accurate to 3 significant figures. (Justify your answer.)

Exercise 18.3b

Use Euler's method to solve the second-order differential equation

$$\ddot{x} = 1 - 2x(\dot{x})^2$$

with initial conditions $x(0) = -1$, $\dot{x}(0) = 0$. Plot x and $\dot{x}$ as functions of t for $0 \le t \le 10$.

18.4 Second-order Runge–Kutta

Euler's method is first order, meaning its errors are proportional to N^{-1}. The error is (approximately) cut in half when we double the resolution. This is not particularly efficient. It generally takes a large number of timesteps N to achieve a modest amount of accuracy. There are other numerical algorithms that are more efficient.

Second-order Runge–Kutta (RK2) is a second-order method, with errors proportional to N^{-2}. With RK2, the error is (approximately) reduced by a factor of 1/4 when the resolution is doubled. Here is the RK2 method:

$$u_h = u_i + F(u_i, t_i)\Delta t/2, \tag{18.9a}$$

$$t_h = t_i + \Delta t/2, \tag{18.9b}$$

$$u_{i+1} = u_i + F(u_h, t_h)\Delta t. \tag{18.9c}$$

In the first equation (18.9a), Euler's method is used to approximate the value of u at the *half* timestep $t_h \equiv t_i + \Delta t/2$. The values of the dependent variables at the half timestep are denoted u_h. The third equation (18.9c) looks just like Euler's method, but instead of using u_i and t_i in the function F, we use the half-timestep values u_h and t_h.

The second-order Runge–Kutta method is not difficult to implement in a numerical code. After importing NumPy and defining parameters, create an array for the discrete times:

```
t = np.linspace(tinitial,tfinal,N+1)
```

Next, define an array for each of the dependent variables using (for example) `np.zeros(N+1)`. The RK2 Eqs. (18.9) should appear in a loop, such as

```
for i in range(N):
```

This implements the N timesteps.

Comments:

- Remember, Eq. (18.9a) is a *set* of equations, one for each of the dependent variables. For example, for the system (18.1), Eq. (18.9a) represents

```
x[i+1] = x[i] + (x[i]*y[i] - z[i] + t[i])*Deltat/2
y[i+1] = y[i] + (2*z[i] + x[i])*Deltat/2
z[i+1] = z[i] + (t[i]**2 - y[i]*z[i])*Deltat/2
```

  Likewise, Eq. (18.9c) is a set of equations. All of the equations (18.9) should be included in the single `for` loop.
- Be careful to implement Eq. (18.9c) properly. The function $F(u_h, t_h)$ is evaluated at the half-timestep values u_h and t_h, but the leading term on the right-hand side is u_i, not u_h.
- The variables u_h and t_h do not need to be defined as arrays. Their values can be reassigned with each cycle through the `for` loop.

Since the error for second-order Runge–Kutta is proportional to N^{-2}, we have $|\text{error}| = CN^{-2}$ for some positive constant C. Take the logarithm of this relation to obtain

$$\log(|\text{error}|) = \log(C) - 2\log(N). \tag{18.10}$$

A plot of $\log(|\text{error}|)$ versus $\log(N)$ should be (approximately) a straight line with slope -2.

Exercise 18.4a

Use RK2 to numerically solve the system of ODE's

$$\dot{x} = -y,$$
$$\dot{y} = x,$$

with initial conditions $x(0) = 1$, $y(0) = 0$, over the time interval $0 \leq t \leq 10$. Carry out a convergence test: Use the exact solution $x(t) = \cos t$, $y(t) = \sin t$ to compute the error in $x(10)$ for various numbers of timesteps. Plot $\log(|\text{error}|)$ versus $\log(N)$ and show that as N becomes large the slope approaches -2.

Second-order Runge–Kutta is better than Euler's method because, most often, it requires fewer timesteps and less compute time to achieve a comparable level of accuracy.

Exercise 18.4b

Use RK2 to solve the system

$$\dot{x} = \sin(tx)$$
$$\dot{y} = y^2 - x$$

with initial conditions $x(0) = 1.0$ and $y(0) = 0.0$, for $0 \leq t \leq 10$. Find the value of $x(10)$ accurate to 5 significant figures. About how many timesteps are required to achieve this level of accuracy? Compare with the number of timesteps required for Euler's method.

18.5 Fourth-order Runge–Kutta

Runge–Kutta methods are a family of numerical algorithms for solving systems of ODEs. The most popular member of this family is fourth-order Runge–Kutta (RK4). Applied to the system $du/dt = F(u, t)$, the RK4 algorithm is

$$u_a = u_i + F(u_i, t_i)\Delta t/2, \tag{18.12a}$$
$$u_b = u_i + F(u_a, t_h)\Delta t/2, \tag{18.12b}$$

$$u_c = u_i + F(u_b, t_h)\Delta t, \tag{18.12c}$$

$$u_d = u_i + F(u_c, t_{i+1})\Delta t, \tag{18.12d}$$

$$u_{i+1} = \frac{1}{3}(u_a + 2u_b + u_c + u_d/2) - \frac{1}{2}u_i. \tag{18.12e}$$

As with RK2, $t_h \equiv t_i + \Delta t/2$ is the time half-way between t_i and t_{i+1}. The variables u_a, u_b, u_c, and u_d are intermediate values of $u(t)$ that are used to evaluate the right-hand side function $F(u,t)$.

Although it is far from obvious, RK4 is a *fourth-order* scheme; that is, the errors are proportional to N^{-4}. A plot of $\log(|\text{error}|)$ versus $\log(N)$ should be (approximately) a straight line with slope -4.

Exercise 18.5a

Use RK4 to solve the system of ODEs

$$\dot{x} = \ln y + \sin x,$$

$$\dot{y} = \ln x + \cos y,$$

with initial conditions $x(0) = y(0) = 1$. Plot graphs of x and y versus t. Find $x(10)$ and $y(10)$ to 6 significant figures.

Exercise 18.5b

Use RK4 to solve the system of ODEs

$$\dot{x} = y,$$

$$\dot{y} = x,$$

with initial conditions $x(0) = 1$, $y(0) = 0$ over the time interval $0 \le t \le 2$. Carry out a convergence test: Use the exact solution $x(t) = \cosh t$, $y(t) = \sinh t$, to compute the errors in $x(2)$ and $y(2)$ for various numbers of timesteps N. Plot a graph of $\log(|\text{error}|)$ versus $\log(N)$ and show that as N becomes large the slope approaches -4.

18.6 `solve_ivp()`

The function `solve_ivp()` from the `scipy.integrate` library solves systems of first-order ordinary differential equations. The acronym "ivp" stands for *initial value problem*. This refers to the types of problems we have been solving all along—a system of differential equations with initial conditions. See the discussion in Sec. 20.1.

Consider the initial value problem

$$\ddot{x} + ax - (\dot{x})^2/x = 0, \tag{18.13}$$

with $x(0) = 1$, $\dot{x}(0) = 0$. The equivalent system of first-order equations is

$$\dot{x} = y, \tag{18.14a}$$

$$\dot{y} = -ax + y^2/x, \tag{18.14b}$$

with $x(0) = 1$ and $y(0) = 0$. The following code solves this system numerically for times $0.0 \leq t \leq 8.0$ and parameter value $a = 0.1$:

```
import numpy as np
import scipy.integrate as si
import matplotlib.pyplot as plt

# right-hand sides of ODEs
def F(t,variables,a):
    x,y = variables
    dxdt = y
    dydt = -a*x + y**2/x
    return dxdt, dydt

# parameters
a = 0.1
ti = 0.0     # initial time
tf = 8.0     # final time

# initial conditions
xi = 1.0
yi = 0.0

# solve the ODEs
u = si.solve_ivp(F,[ti,tf],[xi,yi],args=[a])
t = u.t         # independent variable
```

```
x = u.y[0]    # first dependent variable
y = u.y[1]    # second dependent variable

# plot results
plt.close()
plt.plot(t,x,'r',t,y,'b')
plt.show()
```

Take a close look at the arguments of `solve_ivp()`.

- The first argument `F` is the name of the right–hand sides function.
- The second argument `[ti,tf]` is a list containing the initial and final times.
- The third argument `[xi,yi]` is a list of initial conditions.
- The fourth argument `args=[a]` is an optional list of parameters to pass to the function definition.

Also note the following:

- `solve_ivp()` returns the results as a dictionary, which we call `u`. Discrete values of the independent variable are stored in `u.t`. The corresponding values of the dependent variables are stored in `u.y`. Thus, the first dependent variable is `u.y[0]`, the second dependent variable is `u.y[1]`, etc.
- In the function definition, the line `x,y = variables` assigns the dependent variables to the more familiar names `x` and `y`.

Exercise 18.6a

Run the code. How does the graph look? What are the values of x and y at the final time?

By default, `solve_ivp()` only saves the data at a few discrete times, so the resulting graph is not very smooth. This has nothing to do with the accuracy of the code. You can tell `solve_ivp()` to keep more data points by specifying the option `t_eval`. For example,

```
tee = np.linspace(ti,tf,101)
u = si.solve_ivp(F,[ti,tf],[xi,yi],args=[a],t_eval=tee)
```

instructs `solve_ivp()` to keep `101` data points beginning with the initial time `ti` and ending with the final time `tf`.

Exercise 18.6b

Specify the option `t_eval` with `101` data points. Now how does the graph look? Have the values of x and y at the final time changed? Do x and y change if you replace `101` with `11`?

Exercise 18.6c

Use `solve_ivp()` to solve

$$\dot{x} = \sqrt{x} - ax/y,$$
$$\dot{y} = t/b - y/x,$$

with $a = 1.3$, $b = 2.0$, and initial conditions $x(0) = 3$, $y(0) = 1$. Plot x and y as functions of t for $0 \le t \le 10$.

18.7 ODE solvers with `solve_ivp()`

By default, `solve_ivp()` uses the "RK45" method to solve ODEs. This method uses a combination of fourth- and fifth-order Runge–Kutta algorithms to solve the ODEs, monitor the error, and adjust the timestep as needed. The `solve_ivp()` function can access other solvers, including "RK23" (second- and third-order Runge–Kutta), "DOP853" (eighth-order Runge–Kutta), "Radau" (an order 5 Runge–Kutta type method), and others. You can specify the solver to use by adding an option such as `method = 'DOP853'` as an argument in `solve_ivp()`.

Exercise 18.7a

Solve the system (18.14) using various ODE solvers in `solve_ivp()`. Is there a difference in the results?

How accurate is `solve_ivp()`? If we ask Python to print out the value of a dependent variable at any particular time, it gives us the answer to 16 or 17 significant figures. Should we trust all of the digits? It can be difficult to judge the accuracy of a "canned" solver like those in `solve_ivp()`. This is the main reason why research scientists often prefer to write their own numerical routines.

For the system (18.14), the exact solution is known:

$$x(t) = e^{-at^2/2}, \tag{18.15a}$$

$$y(t) = -ate^{-at^2/2}. \tag{18.15b}$$

In this case, we can determine the accuracy of the ODE solvers by comparing with the exact results. Keep in mind that the accuracy of any given algorithm is highly problem-dependent.

Exercise 18.7b

Solve Eqs. (18.14) using your RK4 code and compute the errors in the numerical values of $x(8)$ and $y(8)$. How many decimal places are correct? Does RK45 do any better? How about DOP853?

Exercise 18.7c

Use your own RK4 code to solve

$$\dot{x} = \sin^2(tx)$$
$$\dot{y} = y^2 - x$$

with initial conditions $x(0) = 1.0$ and $y(0) = 1.0$, for $0 \leq t \leq 10$. Plot the results. Solve the same system using `solve_ivp()` with the RK45 method. Do your code and RK45 agree? Compare using both low and high resolution with your RK4 code.

There are optional arguments that can be given to `solve_ivp()` to increase the accuracy of RK45, DOP853, etc. These options are discussed in Sec. 19.6.

Chapter 19

Driven Damped Pendulum and Chaos

19.1 Equation of motion

The driven, damped pendulum is a relatively simple system that exhibits a wide range of interesting dynamical behaviors, including chaos.[1] There is no analytical solution for this system. It can only be investigated numerically or experimentally.

Figure 19.1 shows a plane pendulum with cord length ℓ and mass m. The angle between the pendulum cord and the vertical is Θ. The pendulum bob is subject to four forces: tension T from the cord, gravity mg, drag F_{drag} due to friction and/or air resistance, and a driving force F_{drive}. The motion is determined from Newton's second law.

Since the motion of the pendulum bob is always perpendicular to the cord, we only need to consider the perpendicular component of Newton's second law. The sum of forces perpendicular to the cord is

$$\sum F_{\perp} = F_{\text{drive}} + F_{\text{drag}} - mg\sin\Theta, \tag{19.1}$$

where $mg\sin\Theta$ is the perpendicular component of the gravitational force. By Newton's second law, $\sum F_{\perp}$ equals the product of mass m and acceleration. Recall that the velocity of an object moving on

[1] A readable account of the driven damped pendulum and chaos can be found in Taylor, J.R. (2005), *Classical Mechanics* (University Science Books).

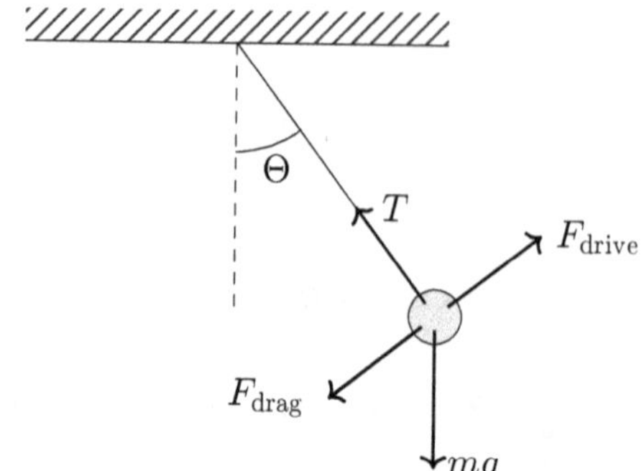

Fig. 19.1. A pendulum with mass m and cord length ℓ. The pendulum makes an angle Θ with respect to the vertical.

a circular path, like the pendulum bob, is $\ell\dot{\Theta}$ where the dot denotes a time derivative. Thus, the acceleration is $\ell\ddot{\Theta}$ and we have

$$m\ell\ddot{\Theta} = F_{\text{drive}} + F_{\text{drag}} - mg\sin\Theta. \tag{19.2}$$

This is a second-order ordinary differential equation for Θ as a function of time t.

Let's assume the drag force is a linear function of the velocity $\ell\dot{\Theta}$, and that it opposes the motion. Thus,

$$F_{\text{drag}} = -\beta\dot{\Theta}, \tag{19.3}$$

where β is some positive constant. Let the external driving force be a periodic function of time,

$$F_{\text{drive}} = \gamma\cos(2\pi t/\tau), \tag{19.4}$$

where γ is a constant amplitude and τ (also a constant) is the time period of the force. Then the differential equation for the driven, dampled pendulum becomes

$$\ddot{\Theta} = -(g/\ell)\sin\Theta - \beta/(m\ell)\dot{\Theta} + \gamma/(m\ell)\cos(2\pi t/\tau). \tag{19.5}$$

The degree of damping is controlled by β, the strength of the driving force is controlled by γ, and τ is the period of the driving force.

Exercise 19.1

Write the second-order differential equation (19.5) as two first-order equations by introducing the new variable $\Omega = \dot{\Theta}$.

19.2 Long-term motion

Imagine pulling the pendulum aside to some initial angle $\Theta(0)$ and releasing it with some initial angular velocity $\Omega(0) = \dot{\Theta}(0)$. The pendulum will swing back and forth. If $\gamma = 0$, so the driving force is turned off, then damping will cause the pendulum to slow down and eventually come to rest at $\Theta = 0$. (Or equivalently, at some multiple of 2π.) This will happen for any initial conditions $\Theta(0)$ and $\Omega(0)$. If $\gamma \neq 0$, we would expect that in the long run, the pendulum's motion will be dictated by the driving force. That is, initially the pendulum's motion will depend on the particular values of $\Theta(0)$ and $\Omega(0)$, but at late times the pendulum's motion will be independent of initial conditions. It will move in some definite manner determined entirely by the driving force. The exact details of this long-term motion will depend on the parameters of the system.

Exercise 19.2

Use RK4 to solve for the motion of the driven, damped pendulum. Choose the parameter values (in SI units) $g = 9.8$, $m = 0.5$, $\ell = 0.11$, $\beta = 0.25$. Also let $\tau = 1.0$ and $\gamma = 1.0$. Carry out two simulations using very different initial conditions for $0 \leq t \leq 20$. On the same graph, plot Θ versus t for both simulations. What are the period and amplitude for t greater than (roughly) 2 or 3? What are the period and amplitude with $\tau = 1.0$ and $\gamma = 0.5$? With $\tau = 2.0$ and $\gamma = 1.0$?

Your results should confirm that after a few drive periods the motion of the pendulum settles into regular, nearly sinusoidal oscillations about $\Theta = 0$ (or some multiple of 2π) with period τ. This long-term solution is called an *attractor*. For the sets of parameters we have investigated thus far, there is unique attractor. The pendulum will aways evolve toward this attractor, independent of initial conditions.

19.3 Code tests

The driven damped pendulum and other chaotic systems can be very difficult to simulate. We need to make sure our results are reliable.

If we had an analytical solution for the system, we could carry out a convergence test: (1) compute the error at various resolutions N; (2) plot $\log|\text{error}|$ versus $\log(N)$; (3) find the slope. Since the errors for RK4 are proportional to N^{-4}, the slope should be approximately -4. This would confirm that the truncation errors are decreasing and the numerical results are becoming more and more precise as N is increased.

Unfortunately, we don't have an analytical solution. But we can carry out a three-point convergence test as discussed in Sec. 17.5 in the context of Euler's method. Here's the idea: Let $\Theta_{(N)}$ denote the final value of Θ (the value at the final time) obtained from a simulation with N timesteps. Since the errors in RK4 are proportional to N^{-4}, we have

$$\Theta_{(N)} = \Theta_{(\text{exact})} + C/(N)^4, \tag{19.6a}$$

$$\Theta_{(2N)} = \Theta_{(\text{exact})} + C/(2N)^4, \tag{19.6b}$$

$$\Theta_{(4N)} = \Theta_{(\text{exact})} + C/(4N)^4. \tag{19.6c}$$

Combine these relations to obtain

$$\frac{\Theta_{(N)} - \Theta_{(2N)}}{\Theta_{(2N)} - \Theta_{(4N)}} = 2^4. \tag{19.7}$$

This is an approximate result, because Eqs. (19.6) aren't exact; they omit terms proportional to higher powers of $1/N$. However, in the limit as N becomes large, the ratio on the left of Eq. (19.7) should approach the value $2^4 = 16$.

Exercise 19.3a

Continue with parameter values $g = 9.8$, $m = 0.5$, $\ell = 0.11$, $\beta = 0.25$, $\tau = 1.0$, and $\gamma = 1.0$ for the driven, damped pendulum. Use a time interval $0 \leq t \leq 10.0$ and any initial conditions. Your code will need to simulate the system at resolutions N, $2N$ and $4N$, then compute the ratio (19.7) using the final values of Θ. Is the ratio close to 16? Does it come closer to 16 as N is increased?

If the ratio approaches 16 for large N, then your results are converging and are probably reliable. Keep in mind, if N is *too* large, then machine roundoff errors can spoil the results.

Another way to check your code is to compare it with another code.

Exercise 19.3b

(Use $g = 9.8$, $m = 0.5$, $\ell = 0.11$, $\beta = 0.25$, $\tau = 1.0$, and $\gamma = 1.0$, and any initial conditions.) Simulate the driven, damped pendulum for times $0 \leq t \leq 10$ using your RK4 code and the built–in function `solve_ivp()` from the `scipy.integrate` library. Experiment with different solvers such as `RK45`, `DOP853`, etc. Plot the results from RK4 and `solve_ivp()` on the same graph. Can you see a difference? How large does N need to be for the plots to look the same?

If all has gone well, your RK4 code should agree with the various `solve_ivp()` solvers. By "agree" we mean that the curves on the graph look visually the same. This is not a definitive test of your code because we don't know how accurate the `solve_ivp()` solvers are. But it should give you some confidence that your code is performing correctly. After all, it is unlikely that your code and the `solve_ivp()` solvers are all wrong in the same way.

19.4 State space

For dynamical systems, the space of coordinates and velocities is called *state space*.[2] Each point in state space represents a unique state of the system. The initial data, in particular, correspond to a single point in state space. As the system evolves in time, its state changes and the initial data point sweeps out a curve through state space.

The periodicity of the motion for the driven, damped pendulum can been seen in a state space plot. Once the pendulum has settled into a nice periodic motion, a plot of Ω versus Θ should show a simple "orbit" that repeats once for each drive period. Figure 19.2 shows three orbits using various values of τ and γ.

You can display the state space orbit by using only the data from late times. For example, you might want to plot data for times that include only the last four drive cycles, $t_f - 4\tau \leq t \leq t_f$, where t_f is the final time. Define the "points per cycle" ppc $= N\tau/t_f$, which is

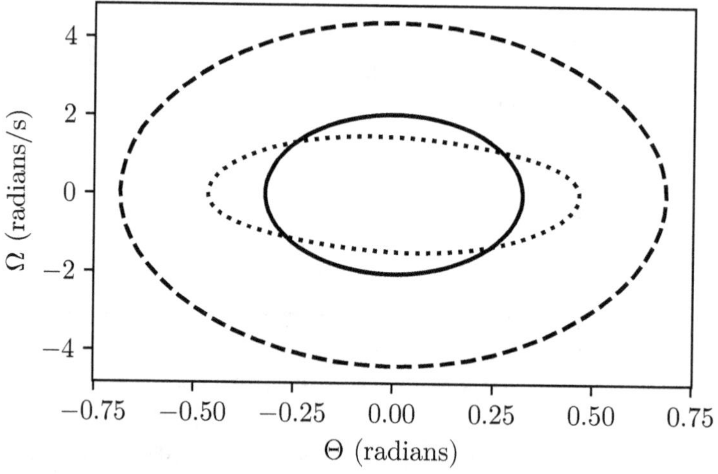

Fig. 19.2. State space orbits for the driven, damped pendulum with driving force parameters $\tau = 1$, $\gamma = 1$ (solid), $\tau = 1$, $\gamma = 2$ (dashed) and $\tau = 2$, $\gamma = 2$ (dotted).

[2]Equivalently, the space of coordinates and momenta is called *phase space*.

the number of data points included in one drive period. (The initial time is 0.) Create a state space plot with

```
ppc = int(N*tau/tfinal)
plt.plot(Theta[N - 4*ppc:N+1],Omega[N - 4*ppc:N+1])
```

(This assumes `matplotlib.pyplot` has been loaded as `plt`.) The array `Theta[N-4*ppc:N+1]` consists of the final `4*ppc` elements of `Theta`, namely, `Theta[N-4*ppc]` through `Theta[N]`. To plot the final `n` drive cycles, use `Theta[N-n*ppc:N+1]`.

Exercise 19.4a

Create a state space plot for the driven, damped pendulum using only the data from late times. (Use $g = 9.8$, $m = 0.5$, $\ell = 0.11$, $\beta = 0.25$, $\tau = 1.0$ and $\gamma = 1.0$.) Start with initial conditions $\Theta(0) = \Omega(0) = 0$. Choose the time interval $0 \leq t \leq 20$, and use enough timesteps to obtain a reliable result. The graph should show a simple closed orbit that repeats once each drive period.

As the driving force amplitude γ is increased, the pendulum exhibits a variety of interesting behaviors.

Exercise 19.4b

Continue using parameter values $g = 9.8$, $m = 0.5$, $\ell = 0.11$, $\beta = 0.25$, and $\tau = 1.0$, initial conditions $\Theta(0) = \Omega(0) = 0$, and time interval $0 \leq t \leq 20$. Let $\gamma = 5.15$. Create a state space plot as well as a plot of Θ versus t. Describe the motion qualitatively. What is the period of the motion?

19.5 Period doubling

Set the initial conditions to $\Theta(0) = -\pi/2$, $\Omega(0) = 0$. Continue to use parameter values $g = 9.8$, $m = 0.5$, $\ell = 0.11$, $\beta = 0.25$, and $\tau = 1.0$.

Exercise 19.5a

Let $\gamma = 5.0$. Create a phase space plot and a plot of Θ versus t. Describe the motion qualitatively. What is the period of the motion?

As the drive amplitude γ is increased beyond about 5.0568, the pendulum exhibits *period doubling.* The pendulum executes a more complicated dynamical behavior with period 2τ, twice the drive period. With further increase in γ the period doubles again, to 4τ. Then 8τ, 16τ, etc. The sequence of period doubling ends at $\gamma \approx 5.1341$. Beyond this threshold, the pendulum motion is not periodic at all—it is chaotic. The range of dynamical behaviors is depicted in Fig. 19.3.

Exercise 19.5b

Choose a value of γ in the middle of the range $5.0568 \lesssim \gamma \lesssim 5.1177$, where the motion of the pendulum has period is 2τ. Be careful that your resolution is sufficiently high to obtain accurate results. You might want to extend your simulation to a final time of $t_{\text{final}} = 100.0$ or more, to make sure you're seeing the long–term behavior. Use state space and Θ versus t plots to interpret the results. If your Python application allows for interactive graphs, it can be helpful to zoom in on small sections of these plots.

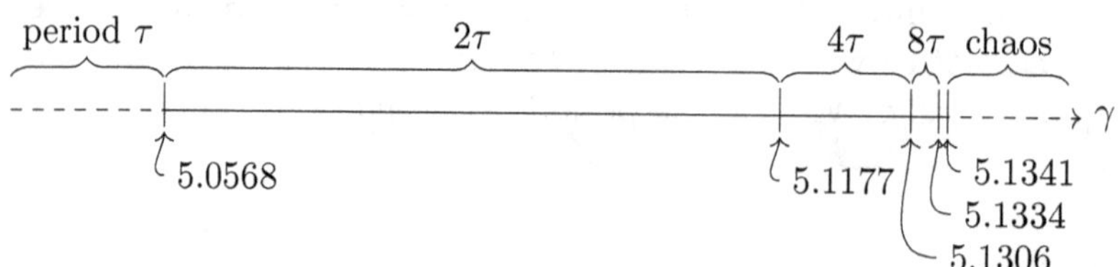

Fig. 19.3. Period doubling. As the drive amplitude γ is increased, the period doubles from τ to 2τ, then 4τ, etc. Motion with period 16τ, 32τ, etc. is squeezed into the narrow range $5.1334 \lesssim \gamma \lesssim 5.1341$. Beyond $\gamma \approx 5.1341$, the motion is chaotic. The threshold values of γ are approximate.

Exercise 19.5c

Choose a value of γ in the middle of the range $5.1177 \lesssim \gamma \lesssim 5.1306$, where the motion of the pendulum has period 4τ. Examine the motion using state space and Θ versus t plots.

These simulations are difficult to carry out for several reasons. As always, simulations must be run at more than one resolution to help insure that truncation error is not affecting the results. If N is too large, machine roundoff errors can be a problem. An additional difficulty is that the ideal steady–state motion (the attractor solution) is never actually reached with a finite run time. As we progress to more complex motion, with period 2τ, 4τ, etc., the periodic behavior takes longer to emerge and is more difficult to identify in a finite run time. Of course we can increase the run time, but then the truncation errors become larger.

The threshold values shown in Fig. 19.3 are approximate. They were obtained from an RK4 code with run time $0 \leq t \leq 400$, at resolution $N = 400000$ and $N = 800000$. To identify the onset of the first period doubling, the values of Θ at late integer times $360, 361, \ldots, 400$ were saved. The code then produces a plot of $\Theta - \Theta_{\rm ave}$, where Θ are the saved values and $\Theta_{\rm ave}$ is the average of these values. If the motion had exactly period $\tau = 1$, the plot would show a horizontal straight line. If the motion had exactly period $2\tau = 2$, the motion would be a zig-zag between two distinct values. Unfortunately, the threshold between period $\tau = 1$ and period $2\tau = 2$ is not distinct, because the motion is not precisely periodic (due to the finite run time). What we actually see is, for $\gamma \lesssim 5.0568$, a zig-zag pattern with decreasing amplitude. If we could run the code for a much longer time, we would expect the amplitude to decrease to zero, indicating a period $\tau = 1$ motion. For $\gamma \gtrsim 5.0568$, the zig-zag pattern is steady, with a constant amplitude. This indicates a true period $2\tau = 2$ motion.

The threshold values between higher period motions were obtained using the same strategy. For example, for the threshold between 2τ and 4τ, late time values of Θ were saved at times separated by period 2, namely, $360, 362, \ldots, 400$. The difference between the saved values and their average was plotted against time. A decreasing amplitude zig-zag pattern indicates motion with period $2\tau = 2$. A constant amplitude zig-zag pattern indicates motion with period $4\tau = 4$.

Table 19.1. The ranges of values for the drive force amplitude γ that yield motion with periods 2τ, 4τ, and 8τ.

Period	Range of γ values
2τ	$5.1177 - 5.0568 = 0.0609$
4τ	$5.1306 - 5.1177 = 0.0129$
8τ	$5.1334 - 5.1306 = 0.0028$

Exercise 19.5d

Use your RK4 code to find the threshold value of γ between period τ and period 2τ. You can use the strategy above. Can you think of a better way to determine the threshold? Do you agree with the result $\gamma = 5.0568$?

You can see from Fig. 19.3 that the threshold values of γ bunch together as the period increases. Table 19.1 shows the range of values of γ for motion with period 2τ, 4τ, and 8τ. (The transition from 8τ to 16τ occurs at $\gamma \approx 5.1334$; this is not shown in Fig. 19.3.) The ratio of successive values, from 2τ to 4τ, is $0.0609/0.0129 = 4.72$. The ratio of successive values from 4τ to 8τ is approximately the same, $0.0129/0.0028 = 4.61$. Careful experiments show that these ratios tend to a fixed value of approximately 4.669. This is called the *Feigenbaum constant.*

What is remarkable about the Feigenbaum constant is that it is *universal.* For a wide class of chaotic systems that pass through a sequence of period doublings on their way to chaos, the ratio of successive ranges of parameter values is given by the Feigenbaum constant.

19.6 Chaos

Keep the same initial data and parameter values as in the previous section.

For values of γ greater than about 5.1341, the driven, damped pendulum exhibits chaos. The late-time motion is chaotic rather than periodic.

Exercise 19.6a

Simulate the driven, damped pendulum over the time interval $0 \le t \le 20$ using RK4. Set the drive amplitude to $\gamma = 5.7$. Describe the motion qualitatively.

Simulating a chaotic system can be tricky. We need to make sure these results are reliable.

Exercise 19.6b

Carry out a three-point convergence test using the RK4 code from the previous exercise. Use the resolutions $N = 25000$, 50000, and 100000 and compute the ratio

$$\frac{\Theta_{(100000)} - \Theta_{(50000)}}{\Theta_{(50000)} - \Theta_{(25000)}} ,$$

where each Θ value is the angle of the pendulum at the final time $t = 20.0$. Is the ratio close to $2^4 = 16$?

You can compare your RK4 results with the built-in ODE solvers from the `scypy.integrate` library. These solvers are accessed with the function `solve_ivp()`, as discussed in Secs. 18.6 and 18.7. The accuracy of the `solve_ivp()` solvers is controlled by two parameters, the absolute tolerance `atol` and the relative tolerance `rtol`. The solvers will adjust the timestep in an attempt to keep the truncation error below `atol + rtol*abs(y)`, where `abs(y)` is the larger of the dependent variables Θ and Ω. The default values for the tolerances are `atol = 10**(-3)` and `rtol = 10**(-6)`. You can change these values to, say, 10^{-8} and 10^{-10} by adding the options `atol = 10**(-8), rtol = 10**(-10)` to the function `solve_ivp()`.

Exercise 19.6c

Simulate the pendulum with RK4 as before. Extend your code to include a simulation using one of the built-in `solve_ivp()` solvers. Plot the results Θ versus t from RK4 and from `solve_ivp()` on the same graph. Do the curves overlap? If not, try adjusting the absolute and relative tolerances in the `solve_ivp()` function. You can also try a different ODE solver.

19.7 Sensitivity to initial conditions

Chaotic systems can be extremely sensitive to initial conditions. You can monitor the sensitivity to initial conditions by creating a version of your code that runs two simulations, one with initial conditions $\Theta(0) = -\pi/2$, $\Omega(0) = 0$ and the other with initial conditions $\Theta(0) = -\pi/2 + \epsilon$, $\Omega(0) = 0$. Compute the difference in angles $\Delta\Theta$ between the two simulations, then plot $\ln|\Delta\Theta(t)|$ versus t.

First, let's look at an example with nonchaotic motion. With $\gamma \lesssim 5.1341$ and $\epsilon = 0.1$, a plot of $\ln|\Delta\Theta(t)|$ versus t will look similar to Fig. 19.4.

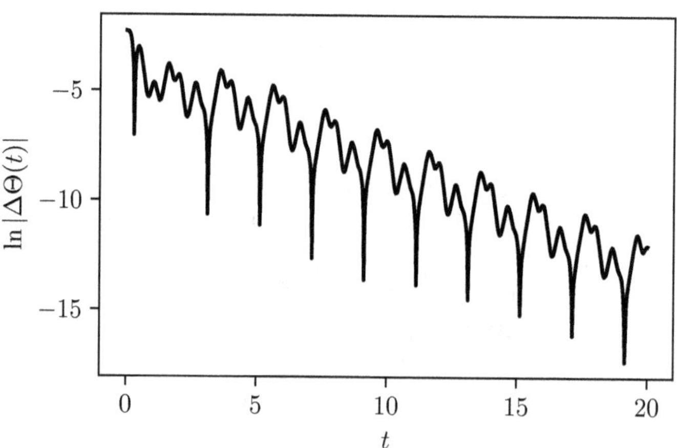

Fig. 19.4. A natural log plot showing the difference in angle for two simulations with different initial conditions. The downward trend shows that the differences decay away exponentially.

Note the downward pointing spikes. These occur when $\Delta\Theta$ changes sign and comes close to zero. If $\Delta\Theta$ happens to be exactly zero at some particular time, Python will complain that it can't evaluate $\ln|\Delta\Theta|$. This is an unlikely occurrence. Most often, because the evolution takes place in discrete time steps, $\Delta\Theta$ will come close to zero as it changes sign but it won't exactly equal zero. The graph produces a sharp downward spike as $\Delta\Theta$ passes close to zero.

We're not actually interested in the spikes. What matters is the upper "envelope" of the curve. This shows a downward trend, decreasing from $\ln(0.1) \approx -2$ to about -12 in the time interval from 0 to 20. The envelope has a slope of roughly $(-12-(-2))/20 = -0.5$. This tells us that the angle difference decreases from its initial value of 0.1 as

$$\Delta\Theta(t) \approx 0.1 \cdot e^{-0.5t}. \tag{19.8}$$

The difference between angles decays away exponentially rapidly.

The slope -0.5 is called the *Lyapunov exponent.* When a Lyapunov exponent is negative, it characterizes the rate at which solutions with different initial conditions evolve toward a common attractor.

Exercise 19.7a

Confirm the results described above. Keep the same parameter values as in the previous sections, and choose a value of γ well below the chaotic regime (say, $\gamma \lesssim 5.1$). What value do you get for the Lyapunov exponent?

When the motion is chaotic, the angle difference increases in time and the Lyapunov exponent is positive. Solutions with different initial conditions rapidly diverge away from one another. The system is sensitive to initial conditions.

Exercise 19.7b

Let $\gamma = 5.7$, so the motion of the pendulum is chaotic. For the second set of initial conditions, choose $\epsilon = 10^{-10}$. Plot $\ln|\Delta\Theta(t)|$ versus t over the time interval $0 \leq t \leq 20$ and find the Lyapunov exponent.

While the angle difference is small, the two solutions will appear nearly identical. However, with the parameters set for chaotic motion, the angle difference eventually approaches a value of order unity and the solutions diverge.

Exercise 19.7c

Keep the parameter values as in the previous exercise, but set the time interval to $0 \leq t \leq 25$. Compare the evolution for the two sets of initial conditions by examining a phase space plot and a plot of Θ versus t. Describe the results.

The motion of a chaotic system like the driven, damped pendulum is deterministic. That is, for any initial point in state space, the differential equations define a unique phase space trajectory. Nevertheless, the motion of a chaotic system is unpredictable in the following sense: We would need to know the initial conditions to arbitrarily high accuracy to predict the exact motion for any extended time.

Why are chaotic systems so difficult to simulate numerically? The difficulty is that any numerical simulation is subject to errors. These errors depend on the numerical algorithm, the resolution, and the limits of machine precision. Any error, whether it occurs in the initial data or at a later time, changes the state of the system. An error at time t "bumps" the system point in phase space from one trajectory to another. If the two trajectories diverge exponentially, this error eventually grows to order unity and drastically alters the results.

One source of error is truncation error, which depends on the numerical algorithm and the resolution.

Exercise 19.7d

Keep the parameter values set to chaotic motion, as in the previous exercise, and initial conditions $\Theta(0) = -\pi/2$, $\Omega(0) = 0$. Set the final time to 100 or more. Compare results using your RK4 code at different resolution values N. Also compare with various `solve_ivp()` solvers, and different settings for the relative and absolute tolerances. Do you trust the results from any of these simulations?

You can reduce the truncation error by increasing the resolution. However, there is a limit to how much the errors can be reduced, due to machine roundoff errors.

Exercise 19.7e

Modify your RK4 code from the previous exercise to run two or more simulations "side-by-side." Compare different ways of writing the equations of motion. For example, you could write the second term in Eq. (19.5) as either `beta/(m*l)*Omega` or `beta*Omega/(m*l)`. Or change the order of the three terms in Eq. (19.5). Why should these changes affect the results? Can you make the differences disappear by increasing the resolution?

Chapter 20

Boundary Value Problems

20.1 The hanging cord

What is the shape of a hanging cord? We can make this question precise. Suspend a cord (or rope or chain) of length L from its ends. The ends are fixed to the points (x_a, y_a) and (x_b, y_b) in the x–y plane, where x is horizontal and y is vertical. What is the curve $y(x)$ taken by the cord?

To answer this question, we first derive an ordinary differential equation satisfied by the function $y(x)$. The equation can be solved numerically as a *boundary value problem*. With a boundary value problem, the freely chosen data are specified by *boundary conditions*. Boundary conditions are specified at the boundaries (or endpoints) of the system. For the hanging cord, the boundary data consist of the numerical values y_a and y_b at the endpoints x_a and x_b.

In contrast, for an *initial value problem* the freely chosen data are given at the initial time. In the previous chapters on differential equations we only considered initial value problems. For dynamical systems such as the driven, damped pendulum of Ch. 18, the initial data are typically the initial positions and initial velocities.

20.2 Differential equation for the cord

Divide the cord into segments of length $\Delta\ell$. Figure 20.1 shows one of these segments. The forces $\vec{F}_L$ and $\vec{F}_R$ on the segment come from the tension in the cord. The force $\vec{F}_L$ acts on the left (L) end of the

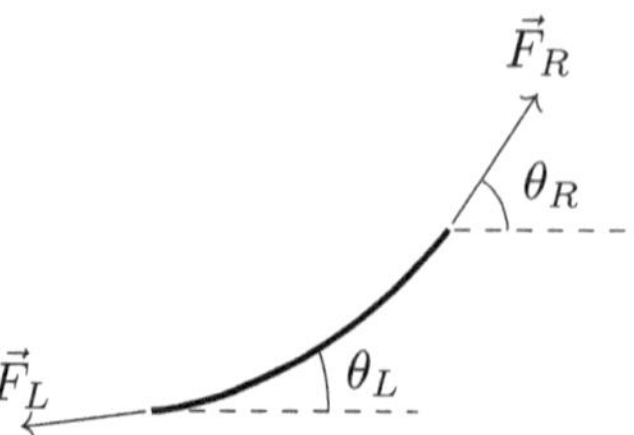

Fig. 20.1. Each segment of cord is acted upon by three forces; tension on the right, tension on the left, and the downward gravitational force (not shown).

segment, and the force $\vec{F}_R$ acts on the right (R) end of the segment. There is also a downward force due to gravity, Δmg, where Δm is the mass of the segment.

Since the segment of cord is in static equilibrium, the sum of forces must vanish. Thus,

$$\vec{F}_L + \vec{F}_R - \Delta mg\hat{y} = 0, \tag{20.1}$$

where $\hat{y}$ is the unit vector pointing upward. The force on the right end of the segment can be split into x (horizontal) and y (vertical) components:

$$\vec{F}_R = T_R \cos\theta_R\,\hat{x} + T_R \sin\theta_R\,\hat{y}. \tag{20.2}$$

Here, T_R is the tension in the cord at the right end of the segment and θ_R is the angle between the cord and the $\hat{x}$ direction at the right end of the segment. Likewise, for the left end of the segment,

$$\vec{F}_L = -T_L \cos\theta_L\,\hat{x} - T_L \sin\theta_L\,\hat{y}. \tag{20.3}$$

Putting these results together we see that the vector equation (20.1) yields

$$T_R \cos\theta_R - T_L \cos\theta_L = 0, \tag{20.4a}$$

$$T_R \sin\theta_R - T_L \sin\theta_L = \Delta mg, \tag{20.4b}$$

for the x and y components.

Consider the first of these results. Since the left and right ends of the segment can be chosen as any two points along the cord, this equation tells us that the product $T\cos\theta$ is constant throughout the cord. Although the tension T and angle θ can vary along the cord, we have

$$T\cos\theta = \tau_x, \tag{20.5}$$

where τ_x is a constant. This constant τ_x is the x-component of the force due to tension.

Now consider the second equation. This can be rewritten as

$$T_R\cos\theta_R\tan\theta_R - T_L\cos\theta_L\tan\theta_L = \Delta mg. \tag{20.6}$$

Since $T\cos\theta$ equals τ_x for both the left and right ends of the segment, we have

$$\tan\theta_R - \tan\theta_L = \Delta mg/\tau_x. \tag{20.7}$$

This can be written more compactly as

$$\Delta(\tan\theta) = \Delta mg/\tau_x, \tag{20.8}$$

where $\Delta(\tan\theta) = \tan\theta_R - \tan\theta_L$ is the change in $\tan\theta$ across the segment.

Let μ denote the mass per unit length of the cord. The mass of the segment is $\Delta m = \mu\Delta\ell$, where $\Delta\ell$ is the segment's length. In the limit as $\Delta\ell$ goes to zero, we can approximate the segment as a straight line and the Pythagorean theorem gives

$$\Delta\ell = \sqrt{\Delta x^2 + \Delta y^2} = \Delta x\sqrt{1 + (\Delta y/\Delta x)^2}. \tag{20.9}$$

Inserting these results into Eq. (20.8) yields

$$\frac{\Delta(\tan\theta)}{\Delta x} = (\mu g/\tau_x)\sqrt{1 + (\Delta y/\Delta x)^2}. \tag{20.10}$$

In the limit $\Delta\ell \to 0$, $\Delta y/\Delta x$ is just the derivative dy/dx. Likewise, $\Delta(\tan\theta)/\Delta x$ is the derivative $d(\tan\theta)/dx$. Recall that at any point on the curve, $\tan\theta$ is the slope at that point: $\tan\theta = dy/dx$. Therefore, $d(\tan\theta)/dx = d^2y/dx^2$ and Eq. (20.10) becomes

$$\frac{d^2y}{dx^2} = (\mu g/\tau_x)\sqrt{1 + (dy/dx)^2}. \tag{20.11}$$

This ordinary differential equation describes the shape of a hanging cord.

The general solution of Eq. (20.11) is

$$y(x) = a + \frac{\cosh((\mu g/\tau_x)x + b)}{(\mu g/\tau_x)}. \tag{20.12}$$

This curve is called a *catenary*. The constants a and b are determined by the boundary conditions.

Exercise 20.2

Verify that the catenary (20.12) satisfies the differential equation (20.11). Then let $g = 9.8\,\mathrm{m/s^2}$, $\mu = 0.1\,\mathrm{kg/m}$, $\tau_x = 0.3\,\mathrm{kg \cdot m/s^2}$ and use the boundary conditions $y(0) = 0.0\,\mathrm{m}$, $y(1) = 0.5\,\mathrm{m}$ to determine the constants a and b. Plot $y(x)$ versus x. (Suggestion: Use SymPy.)

20.3 Relaxation

Equation (20.11) defines a boundary value problem for the shape of a hanging cord. The boundary values are the values of y at the endpoints x_a and x_b. A simple numerical technique for solving such a problem is called *relaxation*. Divide the x-axis into N equal subintervals, each of width Δx, as shown in Fig. 20.2. The x-coordinate values of the nodes are x_i where $i = 0, \ldots, N$. The endpoints are $x_0 = x_a$ and $x_N = x_b$. The y-coordinate value of the cord at node x_i is denoted y_i.

We begin by discretizing the catenary equation Eq. (20.11). In particular, we need discrete approximations for the derivatives

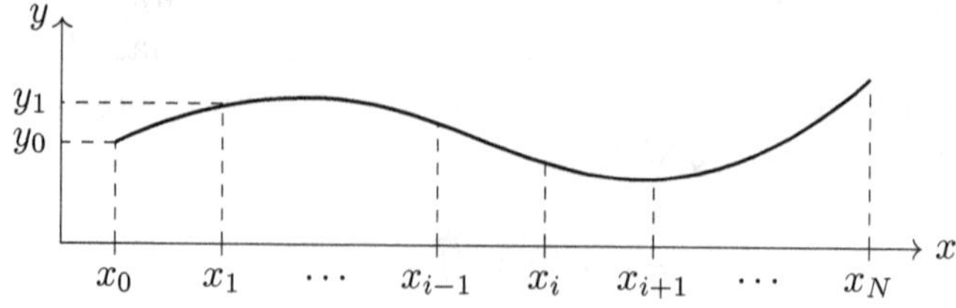

Fig. 20.2. The discrete grid has nodes x_i for $i = 0, \ldots, N$. The endpoints are $x_0 = x_a$ and $x_N = x_b$. The separation between adjacent nodes is Δx.

$y' = dy/dx$ and $y'' = d^2y/dx^2$. The three-point centered difference approximations for the first and second derivatives at node x_i are

$$y'(x_i) \approx \frac{y_{i+1} - y_{i-1}}{2\Delta x}, \tag{20.13a}$$

$$y''(x_i) \approx \frac{y_{i+1} - 2y_i + y_{i-1}}{\Delta x^2}. \tag{20.13b}$$

These are Eqs. (16.18) and (16.21) with a few changes in notation. Inserting these expressions into the catenary equation, we have

$$(y_{i+1} - 2y_i + y_{i-1})/\Delta x^2 = (\mu g/\tau_x)\sqrt{1 + (y_{i+1} - y_{i-1})^2/(2\Delta x)^2}. \tag{20.14}$$

Now solve for y_i:

$$y_i = \frac{1}{2}(y_{i+1} + y_{i-1}) - (\mu g/\tau_x)\frac{\Delta x^2}{2}\sqrt{1 + (y_{i+1} - y_{i-1})^2/(2\Delta x)^2}. \tag{20.15}$$

What can we do with this result? It appears useless. After all, we can't use it to find y_i since we don't know y_{i+1} or y_{i-1}. However, if we have a reasonable approximation for each of the y_i's, we can use this relationship to improve that approximation.

To implement this idea, start with a trial solution. A trial solution is a guess for each value of y_i that satisfies the boundary conditions $y_0 = y_a$ and $y_N = y_b$. We then apply Eq. (20.15) to each of the interior nodes $i = 1, \ldots, N-1$, one after the other. That is, we evaluate the right-hand side of Eq. (20.15) at $i = 1$ and use the result as a new value for y_1. Then evaluate the right-hand side of Eq. (20.15) at $i = 2$ and use the result as a new value for y_2. Continue for each value of i through $N - 1$.

This calculation is called a *relaxation sweep*. The end result of a single relaxation sweep is an improved set of values for $y_1, \ldots, y_{N-1}$. (The values y_0 and y_N do not need to be improved, since they are fixed by the boundary conditions.) We now execute a second relaxation sweep by inserting the improved y_i's into the right-hand side of Eq. (20.15). This process is repeated, as many times as needed, until the values for y_i *relax* to an approximate solution of Eq. (20.14).

The relaxation process is typically slow, and the exact solution of Eq. (20.14) is never actually reached. How many relaxation sweeps should you use? In practice, your code should continue until the discrete equation (20.14) is satisfied to some desired degree of accuracy. We will address this issue in Sec. 20.6.

Keep in mind: Even if we could find the exact solution of the discrete equation (20.14), it would not be an exact solution of the original differential equation (20.11). This is due to truncation errors. The discrete equation is obtained from the differential equation by replacing derivatives with centered difference formulas. These formulas are approximations, with errors proportional to Δx^2.

Exercise 20.3

Write a code to solve the differential equation (20.11). Set $(\mu g/\tau_x) = 5.0$ and choose boundary conditions $y(0) = 0$ and $y(1) = 2$. Use 5000 relaxation sweeps and $N = 101$ nodes. For your trial solution, use the straight line $y_i = 2.0\, x_i$. Have your code plot y versus x after every 1000 sweeps. How much does the value of y at the midpoint, `y[51]`, change between 1000 and 2000 sweeps? Between 2000 and 3000 sweeps? 3000 and 4000? 4000 and 5000?

20.4 Gauss–Seidel and Jacobi methods

You probably wrote your code using a `for` loop, or similar control structure, to sweep across the interior nodes $i = 1, \ldots, N-1$. For example, you might use

```
for i in range(1,N):
    y[i] = 0.5*(y[i+1] + y[i-1]) + ...
```

Let's unwind the `for` loop. The first time through the loop, $i = 1$ and the computer executes the command `y[1] = 0.5*(y[2] + y[0]) + ...`. The right-hand side is evaluated using the trial values for `y[2]` and `y[0]`, then the computer overwrites the trial value for `y[1]` with the right-hand side value. The second time through the loop, $i = 2$ and the computer executes the command `y[2] = 0.5*(y[3] + y[1]) + ...`. The right-hand side is evaluated using the trial value for `y[3]`, and the recently computed new value for `y[1]`. The computer then overwrites the trial value for `y[2]` using the right-hand side value, which is obtained from a mixture of old `y[3]` and new `y[1]`. This mixture of old and new values continues for the remaining cycles through the loop.

Is this really what we want our code to do? An alternative to the code segment above is

```
yold = np.copy(y)
for i in range(1,N):
    y[i] = 0.5*(yold[i+1] + yold[i-1]) + ...
```

In this case, the computer always uses "old" values on the right-hand side, never a mixture of old and new.

It turns out that both approaches are fine. However, the first approach, using a mixture of old and new y values, is easier to implement and is usually faster. This approach is called the *Gauss–Seidel method.* The second approach, which uses only old values for the right-hand side evaluation, is *Jacobi's method.*

Exercise 20.4a

Compare the Gauss–Seidel and Jacobi methods for the hanging cord. In each case, consider the midpoint value of y after 5000 sweeps. For which option does the solution relax most quickly?

Exercise 20.4b

Try a third approach. For each sweep, first step through the odd values of i. This will update the odd `y[i]`'s to new values. Then step through the even values of i, using the updated `y[i]`'s on the right-hand side. How does this approach compare to the Gauss–Seidel and Jacobi methods?

20.5 Speeding up your code

Here is some practical advise. With NumPy's *intrinsic indexing*, the code for Jacobi's method

```
yold = np.copy(y)
for i in range(1,N):
    y[i] = 0.5*(yold[i+1] + yold[i-1]) + ...
```

can be replaced by the single statement

```
y[1:N] = 0.5*(y[2:N+1] + y[0:N-1]) + ...
```

The syntax `y[1:N]` (which can also be written as `y[1:-1]`) denotes the range of elements beginning with `y[1]` and ending with `y[N-1]`. The syntax `y[2:N+1]` (which can also be written as `y[2:]`) denotes the range of elements beginning with `y[2]` and ending with `y[N]`. The syntax `y[0:N-1]` (which can also be written as `y[:-2]`) denotes the range of elements beginning with `y[0]` and ending with `y[N-2]`. Your Jacobi's code will run faster if you use the single statement rather than an explicit `for` loop.

Exercise 20.5a

Solve the hanging cord problem using Jacobi's method with intrinsic indexing. Use the `time()` function from the `time` library to measure the execution time for your code. How much faster does your code run with intrinsic indexing? How does Jacobi's method with intrinsic indexing and 10000 sweeps compare to the Gauss–Seidel method with 5000 sweeps?

Another strategy for speeding up the solution process is called *successive overrelaxation* (SOR). It is a variant of the Gauss–Seidel method. Replace the finite difference formula (20.15) with the equivalent expression

$$y_i = (1-\omega)y_i + \omega\left[\frac{1}{2}(y_{i+1} + y_{i-1}) \right.$$
$$\left. - (\mu g/\tau_x)\frac{\Delta x^2}{2}\sqrt{1 + (y_{i+1} - y_{i-1})^2/(2\Delta x)^2}\right]. \tag{20.16}$$

For each relaxation sweep, evaluate the right-hand side using the Gauss–Seidel method (a mixture of old and new values of y_{i-1}, y_i, and y_{i+1}). The result becomes the new value of y_i.

For $\omega = 1$, the SOR expression (20.16) is identical to the original finite difference formula. For ω larger than 1, but not too large, the relaxation process is more efficient; that is, the numerical solution relaxes more quickly. You can see how quickly the solution is relaxing by monitoring the change in the midpoint value `y[51]` from, say, 4000 sweeps to 5000 sweeps.

Unfortunately, if ω is too large, the SOR algorithm will fail. Determining an appropriate value of ω is a matter of trial and error.

Exercise 20.5b

Apply SOR to the hanging cord problem. Experiment with different values of ω. Does SOR improve the efficiency of your Gauss–Seidel code? What happens if ω is too large? What is the (approximate) maximum value of ω?

20.6 Residual

How many relaxation sweeps should you use? In Sec. 20.3, we noted that the sweeps should continue until the numerical solution satisfies the discrete equation (20.14) to some desired degree of accuracy. We can judge the accuracy of the numerical solution by defining the *residual* at each interior node $i = 1, \ldots, N-1$:

$$\mathcal{R}_i = (y_{i+1} - 2y_i + y_{i-1})/\Delta x^2 - (\mu g/\tau_x)\sqrt{1 + (y_{i+1} - y_{i-1})^2/(2\Delta x)^2}. \tag{20.17}$$

As the numerical solution y_i relaxes towards an exact solution of Eq. (20.14), the absolute value of the residual decreases.

Let's assume that at each interior node i, we want $|\mathcal{R}_i|$ to be less than some small number ϵ. This will be the case if the maximum value of $|\mathcal{R}_i|$ is less than ϵ.

Exercise 20.6

Create a code to solve the hanging cord problem and compute the residual after every, say, 100 relaxation sweeps. (Use either the Gauss–Siedel or Jacobi method.) Have your code print out the values of $\max|R_i|$ and continue carrying out sweeps until $\max|R_i|$ drops below some prescribed value ϵ. How many sweeps are required with $\epsilon = 0.001$? 0.00001?

20.7 Length of the cord

The numerical solution gives us the shape of the cord for some chosen value of the constant $\mu g/\tau_x$. What is the length of the cord? The length of the curve $y(x)$ is

$$L = \int_{x_a}^{x_b} \sqrt{1 + (dy/dx)^2} dx, \tag{20.18}$$

as described in basic calculus. Using the differential equation (20.11), we can write this as

$$L = \frac{\tau_x}{\mu g} \int_{x_a}^{x_b} \left(\frac{d^2 y}{dx^2} \right) dx, \tag{20.19}$$

then integrate to obtain

$$L = \frac{\tau_x}{\mu g} \left(\frac{dy}{dx} \right) \Bigg|_{x_a}^{x_b} . \tag{20.20}$$

The derivatives at the endpoints can be approximated using the two point forward and backward difference formulas (16.4) and (16.6). This gives

$$L \approx \frac{\tau_x}{\mu g \Delta x} (y_N - y_{N-1} - y_1 + y_0). \tag{20.21}$$

Exercise 20.7a

Extend your previous code to include a calculation of the cord length L. Use these results to plot a graph of L versus the constant $\mu g/\tau_x$.

This code produces the cord shape $y(x)$ and the cord length L for a given value of $\mu g/\tau_x$. This isn't really what we want. We would like to specify the length L, rather than the constant $\mu g/\tau_x$, and have the code determine the appropriate shape. Here's one strategy for computing the shape of a hanging cord given its length L and the boundary conditions $y(x_a) = y_a$ and $y(x_b) = y_b$.

Start with a code that computes the cord shape and length for a given constant $\mu g/\tau_x$. For notational simplicity, let $k = \mu g/\tau_x$.

Also let L denote the length that you specify, and ℓ denote the numerically computed length. Extend your code to compute the difference $L - \ell$ for a given k. Now imagine that this entire code is a function $\mathcal{F}(k)$ whose input is k and whose output is $L - \ell$. We can use the bisection method of Sec. 11.3 to find the value of k that satisfies $\mathcal{F}(k) = 0$.

To apply bisection, start with k_L and k_R such that $\mathcal{F}(k_L)$ and $\mathcal{F}(k_R)$ have opposite signs. Next, compute the midpoint $k_M = (k_L + k_R)/2$. If $\mathcal{F}(k_M)$ has the same sign as $\mathcal{F}(k_L)$, replace k_L with k_M. If $\mathcal{F}(k_M)$ has the same sign as $\mathcal{F}(k_R)$, replace k_R with k_M. Repeat this process until $|k_R - k_L| < 2\epsilon$, where ϵ is some chosen tolerance. You can take $(k_L + k_R)/2$ as the final answer for the value of k.

What value should you use for the tolerance ϵ? Keep in mind that the relaxation algorithm contains truncation errors whose size depends on the number of grid points. Errors also arise because the number of sweeps is finite, and the numerical solution is never fully relaxed. It would be foolish to set the tolerance too low. Even if the result for k is highly accurate, the shape of the cord will contain errors from these other sources.

Exercise 20.7b

Find the shape of a hanging cord of length $L = 3.0\,\text{m}$ with boundary values $y(0) = 0.0\,\text{m}$ and $y(1.0) = 2.0\,\text{m}$. You might want to use a function definition for the part of your code that computes $\mathcal{F}(k)$. Use bisection to determine the constant $k = \mu g/\tau_x$, and graph the solution $y(x)$.

20.8 Euler–Bernoulli beam theory

Euler–Bernoulli beam theory describes the deflection of an elastic beam in response to a load (forces), assuming the deflection is small. Let the center of the unloaded beam extend along the x-axis. When a load is applied in the y-direction, the Euler–Bernoulli equation gives the deflection y away from the x-axis as

$$EI\frac{d^4y}{dx^4} = q(x). \tag{20.22}$$

Here, $q(x)$ is the force per unit length, which can vary along the length of the beam. Hence, q is a function of x. The constant E is the elastic modulus (Young's modulus) of the material. The constant I is the second moment of the cross-section of the beam, defined by

$$I = \iint z^2 \, dy \, dz. \tag{20.23}$$

For a square cross-section with sides of length ℓ, we have $I = \ell^4/12$.

The Euler–Bernoulli equation is a fourth-order differential equation. For a given beam, the solution depends on the load $q(x)$ as well as the boundary conditions applied at the ends. There are three basic types of boundary conditions: a simple support

$$y\big|_{end} = \text{const}, \quad y''\big|_{end} = 0; \tag{20.24}$$

a clamped end

$$y\big|_{end} = \text{const}, \quad y'\big|_{end} = \text{const}; \tag{20.25}$$

and a free end

$$y''\big|_{end} = 0, \quad y'''\big|_{end} = 0. \tag{20.26}$$

Each of the constants should be small, since Euler–Bernoulli theory assumes the deflection is small.

The Euler–Bernoulli equation can be solved numerically using relaxation. Begin by replacing the fourth derivative of y with the centered finite difference approximation

$$y''''(x_i) \approx \frac{y_{i-2} - 4y_{i-1} + 6y_i - 4y_{i+1} + y_{i+2}}{\Delta x^4}. \tag{20.27}$$

Then the Euler–Bernoulli equation (20.22) becomes

$$y_i = \frac{1}{6}(-y_{i+2} + 4y_{i+1} + 4y_{i-1} - y_{i-2}) + \frac{q(x_i)}{EI}\frac{\Delta x^4}{6}. \tag{20.28}$$

Begin the solution process with a trial solution $y_i = 0$ for $i = 0, \ldots, N$. For each relaxation sweep, evaluate the right-hand side of Eq. (20.28) using the current values of y_i. This yields new values for the amplitudes at the interior nodes, $y_2, \ldots, y_{N-2}$. New values for y_0, y_1, y_{N-1}, and y_N are obtained from boundary conditions.

To be concrete, let's consider a beam that is cantelievered from the left end. That is, the left end is clamped and the right end is free. The boundary conditions (20.25) at the left end are discretized as

$$y_0 = 0, \qquad \frac{y_1 - y_0}{\Delta x} = 0, \tag{20.29}$$

where the two-point forward difference approximation (16.4) is used in place of the first derivative y'. We can solve these equations for y_0 and y_1:

$$y_0 = 0, \tag{20.30a}$$

$$y_1 = 0. \tag{20.30b}$$

The values for y_0 and y_1 remain unchanged from one relaxation sweep to the next.

The right end of the beam is free, and satisfies the boundary conditions (20.26). These conditions can be discretized as

$$\frac{y_N - 2y_{N-1} + y_{N-2}}{\Delta x^2} = 0, \tag{20.31a}$$

$$\frac{y_N - 3y_{N-1} + 3y_{N-2} - y_{N-3}}{\Delta x^3} = 0, \tag{20.31b}$$

and solved for y_{N-1} and y_N:

$$y_N = 3y_{N-2} - 2y_{N-3}, \tag{20.32a}$$

$$y_{N-1} = 2y_{N-2} - y_{N-3}. \tag{20.32b}$$

As a part of each relaxation sweep, after new values for the interior amplitudes $y_2, \ldots, y_{N-2}$ have been computed, Eqs. (20.32) are used to determine new values for y_N and y_{N-1}.

Exercise 20.8a

Use the methods of Ch. 16 to verify the finite difference formulas (20.27) and (20.31).

Exercise 20.8b

A steel rod has density $\rho = 7850\,\text{kg/m}^3$ and elastic modulus $E = 2.05 \times 10^{11}\,\text{kg}/(\text{m}\cdot\text{s}^2)$. The rod is 2.0 m long and has a square cross-section, $\ell = 0.01\,\text{m}$ on each side. The load is the rod's weight, so the force per unit length is $q(x) = -\rho\ell^2$. The left end of the rod is clamped, the right end is free. Use relaxation to solve the Euler–Bernoulli equation. Monitor the residual. Plot $y(x)$ and determine the deflection at the right end. Hint: SOR is helpful.

Exercise 20.8c

The rod from the previous exercise is now held up with simple supports at each end. Discretize the boundary conditions (20.24) and apply them to your code. Use relaxation to solve the Euler–Bernoulli equation and monitor the residual. Plot $y(x)$. How much does the rod sag in the middle?

Chapter 21

Partial Differential Equations I

Partial differential equations (PDEs) describe systems with more than one independent variable. Many examples involve wave motion, such as water waves, electromagnetic waves, sound waves, etc. In each of these cases, the dependent variable is the amplitude of the wave and the independent variables are the time and space coordinates. Other examples of PDEs include the Schrödinger equation, which describes the quantum mechanical wave function, the Poisson equation, which describes the electrostatic and gravitational potentials, and the heat equation, which describes the flow of thermal energy through a body.

21.1 Waves on a string

Consider a string stretched along the x-axis, under tension T. Perhaps a guitar string or a violin string. The mass per unit length of the string is μ. We will ignore gravity and consider only small amplitude, transverse waves. That is, we only allow small displacements of the string away from the x-axis in the y-direction. Our goal is to describe the displacement amplitude $y(t, x)$ as a function of the two independent variables t and x.

Figure 21.1 shows a short segment of string. The forces on the right and left ends of the segment depend on the tension T and the angles θ_R, θ_L that the ends make with the x-direction. The y component of force on the string segment is

$$F_y = T \sin\theta_R - T \sin\theta_L . \tag{21.1}$$

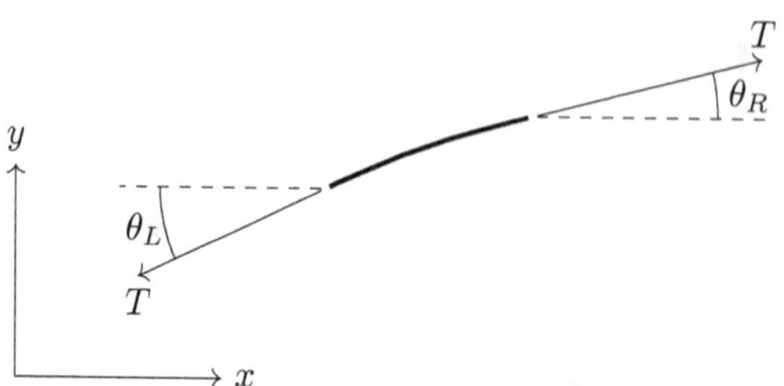

Fig. 21.1. A segment of string under tension T.

By applying Newton's second law, we find

$$\mu \Delta x \frac{d^2 y}{dt^2} = T(\sin\theta_R - \sin\theta_L), \tag{21.2}$$

where Δx is the length of the segment (assuming displacements are small) and $\mu \Delta x$ is the segment's mass. For small displacements of the string the angles θ_R and θ_L will be small. To a good approximation the sin's can be replaced with tan's. But the tangent of θ_R is the slope dy/dx at the right end of the segment. Likewise, the tangent of θ_L is the slope at the left end of the segment. The equation of motion becomes

$$\mu \frac{d^2 y}{dt^2} = \frac{T}{\Delta x}\left(\left.\frac{dy}{dx}\right|_R - \left.\frac{dy}{dx}\right|_L \right). \tag{21.3}$$

In the limit as the segment size shrinks to zero, the right-hand side (excluding the factor T) equals the second derivative of y; that is,

$$\lim_{\Delta x \to 0} \frac{1}{\Delta x}\left(\left.\frac{dy}{dx}\right|_R - \left.\frac{dy}{dx}\right|_L \right) = \frac{d^2 y}{dx^2}. \tag{21.4}$$

Therefore, the PDE that governs small–amplitude wave motion on a string is

$$\frac{d^2 y}{dt^2} = c^2 \frac{d^2 y}{dx^2}, \tag{21.5}$$

where $c^2 = T/\mu$. This is the *wave equation.* Using dots to denote derivatives with respect to t and primes to denote derivatives with respect to x, the wave equation can be written as $\ddot{y} = c^2 y''$.

It is easy to verify that the wave equation has solutions of the form

$$y(t, x) = F(x + ct) + G(x - ct), \tag{21.6}$$

where F and G are arbitrary functions of their arguments. In fact, Eq. (21.6) is the general solution. Any solution of the wave equation can be described as a superposition of a left-moving wave $F(x + ct)$ and a right-moving wave $G(x - ct)$. A purely left-moving wave $y(t, x) = F(x + ct)$ satisfies $y(t + T, x - cT) = y(t, x)$; the waveform shifts to the left by an amount cT during time T. Likewise, a purely right-moving wave $y(t, x) = G(x - ct)$ satisfies $y(t + T, x + cT) = y(t, x)$; the waveform shifts to the right by an amount cT during time T. The constant c is the *wave speed.*

Exercise 21.1

Verify that $y(t, x) = F(x + ct) + G(x - ct)$ satisfies the wave equation for any functions F and G.

21.2 CTCS algorithm

A PDE such as (21.5) can be solved numerically through finite differencing. Let us assume that the ends of the string are held fixed at locations x_a, x_b on the x-axis. Divide the spatial interval $x_a \leq x \leq x_b$ into J segments, each of length $\Delta x = (x_b - x_a)/J$. This defines $J+1$ *spatial nodes* at locations $x_0, x_1, \ldots, x_J$. That is, $x_j = x_a + j\Delta x$, where $j = 0, \ldots, J$. We also divide the time axis into timesteps of length Δt. The *temporal nodes* are denoted $t^0, t^1, \ldots, t^N$. That is, $t^n = n\Delta t$ where $n = 0, \ldots, N$. Let[1]

$$y_j^n \equiv y(t^n, x_j) \tag{21.7}$$

denote the amplitude of the string at spatial location x_j and time t^n. See Fig. 21.2.

[1]The superscripts on t and y are indices, not powers.

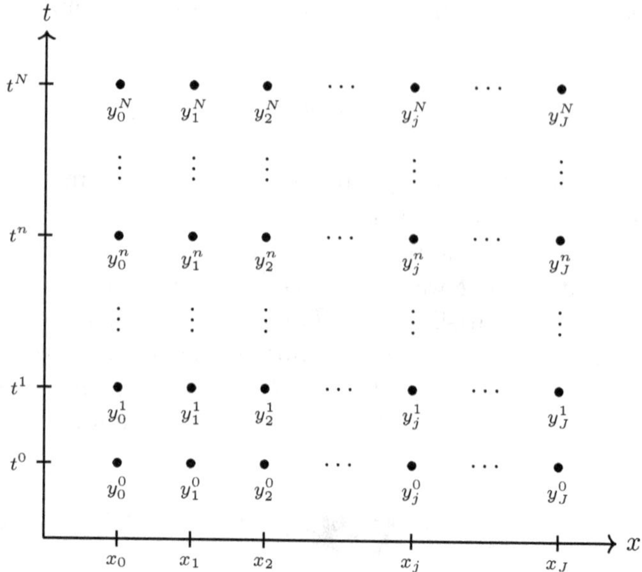

Fig. 21.2. The string lies along the x-axis, which is divided into J segments of length Δx. The $J+1$ spatial nodes are denoted $x_0, \ldots, x_J$, where $x_0 \equiv x_a$ is the left end and $x_J \equiv x_b$ is the right end. The time axis is divided into timesteps of length Δt. The temporal nodes are denoted $t^0, \ldots, t^N$. The amplitude of the string at spatial location x_j and time t^n is denoted y_j^n.

The derivatives of $y(t,x)$, evaluated at node x_j and t^n, can be approximated as

$$\ddot{y}(t^n, x_j) \approx \frac{y_j^{n+1} - 2y_j^n + y_j^{n-1}}{\Delta t^2}, \tag{21.8a}$$

$$y''(t^n, x_j) \approx \frac{y_{j+1}^n - 2y_j^n + y_{j-1}^n}{\Delta x^2}, \tag{21.8b}$$

using the three-point centered difference formula (16.21). Inserting these expressions into the PDE (21.5) and solving for y_j^{n+1}, we find

$$y_j^{n+1} = 2y_j^n - y_j^{n-1} + \left(\frac{c\Delta t}{\Delta x}\right)^2 (y_{j+1}^n - 2y_j^n + y_{j-1}^n). \tag{21.9}$$

The numerical method based on this approximation to the wave equation is called the centered-time, centered-space (CTCS) algorithm.

Here's the idea. Equation (21.9) must hold for each value of the time index n. In particular, for $n = 1$, we have

$$y_j^2 = 2y_j^1 - y_j^0 + \left(\frac{c\Delta t}{\Delta x}\right)^2 (y_{j+1}^1 - 2y_j^1 + y_{j-1}^1). \qquad (21.10)$$

Thus, if we know the amplitudes at each spatial node at times t^0 and t^1, we can use Eq. (21.10) to predict the amplitudes at time t^2 at each of the interior nodes $j = 1, \ldots, J-1$. Since the endpoints of the string are fixed, the amplitudes at the endpoints are simply $y_0^2 = y_0^1$ and $y_J^2 = y_J^1$. Having found all of the amplitudes at time t^2, we can apply Eq. (21.9) with $n = 2$ to obtain the amplitudes at t^3. Similarly, we find the amplitudes at t^4, t^5, etc.

Let's put together a numerical code to solve the wave equation (21.5) using the CTCS algorithm.

- Import NumPy (as `np`) and other libraries.
- Choose values for the parameters including x_a, x_b, and c.
- Set up the numerical grid; for example:

```
J = 100              # number of space intervals
N = 500              # number of timesteps
dx = (xb-xa)/J
dt = 0.2*dx/c
x = np.linspace(xa,xb,J+1)
```

  Here, the timestep `dt` is chosen to be proportional to `dx/c`; we will discuss this later.
- Define arrays for y^{n-1}, y^n and y^{n+1}:

```
ym = np.zeros(J+1)     # ym = "y minus" = y at time t^(n-1)
y = np.zeros(J+1)      # y at time t^n
yp = np.zeros(J+1)     # yp = "y plus"  = y at time t^(n+1)
```

- Choose "initial" data; that is, data at times t^0 and t^1:

```
y = some function of x
yp = some function of x
```

  The array `y` holds the values of y_j^0, and the array `yp` holds the values of y_j^1. This completes the first timestep.
- In preparation for the second timestep, from t^1 to t^2, copy the arrays `y` and `yp` into `ym` and `y`:

```
ym = np.copy(y)
y = np.copy(yp)
```

- Evolve the system forward in time using the CTCS scheme (21.9):

```
for n in range(2,N+1):
    for j in range(1,J):
        yp[j] = 2.0*y[j] - ym[j]\
            + (c*dt/dx)**2 *(y[j+1] - 2.0*y[j] + y[j-1])
    yp[0] = y[0]
    yp[J] = y[J]
    ym = np.copy(y)
    y = np.copy(yp)
```

  The loop over n begins at $n = 2$ because the first timestep was taken when we assigned values to y_j^1. Inside the loop over n, `yp[j]` is computed for all interior nodes $j = 1, \ldots, J-1$. Boundary conditions are used to determine the endpoint values `yp[0]` and `yp[J]`. The arrays `y` and `yp` are then copied to `ym` and `y` in preparation for the next timestep.
- Add code as necessary to plot graphs and print (or save) `y` at various times t^n. For example, to plot $y(x)$ after every 100 timesteps, simply include

```
if (n%100 == 0):
    matplotlib.pyplot.plot(x,y)
```

  inside the loop over n.

Since the wave equation contains second-order derivatives in time, the initial data consist of the initial shape $y(0,x)$ as well as the initial velocity $\dot{y}(0,x)$. For example, let the string start from rest with a "Gaussian pulse" shape

$$y(0,x) = Ae^{-x^2/\sigma^2}, \tag{21.11}$$

where A and σ are constants. You can implement this in your code by setting

$$y_j^0 = Ae^{-x_j^2/\sigma^2} \tag{21.12}$$

for all j. Since the string is initially at rest, $\dot{y}(0,x) = 0$. To implement this condition, replace the time derivative with the two-point forward difference approximation (16.4). This yields $(y_j^1 - y_j^0)/\Delta t = 0$, so that

$$y_j^1 = y_j^0 \tag{21.13}$$

for all j. In the numerical code described above, y_j^0 is placed into the array `y` and y_j^1 is placed into the array `yp`.

Exercise 21.2a

Write a code to solve the wave equation with boundary conditions $y(t, x_a) = y(t, x_b) = \text{const}$ at the endpoints $x_a = -5.0$ and $x_b = 5.0$ (in SI units). Use the CTCS scheme with $J = 100$ or more. Choose the mass density $\mu = 0.01$ and tension $T = 25.0$, which yields a wave speed of $c = 50.0$. Start the string from rest, Eq. (21.13), with a Gaussian shape, Eq. (21.12). Choose any value for the constant A, such as $A = 1$. Choose a value for σ such that the endpoint values y_0^0 and y_J^0 are close to zero. Have your code plot y versus x at various times on the same graph.

Exercise 21.2b

Estimate the location of the peak in $y(t, x)$ at various times, and use this to verify that the wave speed is $c = 50.0$.

You might wonder why we use the one-dimensional arrays `ym[j]`, `y[j]`, and `yp[j]` to represent y_j^{n-1}, y_j^n, and y_j^{n+1}. Why not create a single two-dimensional array `y[n,j]` to hold the string amplitudes for each spatial node x_j at each time t^n? This would be easier to implement numerically. The difficulty is that the array `y[n,j]` would contain $(N+1)(J+1)$ elements. If N and J are large, your computer might run into memory issues. This can be especially problematic for PDEs in two or three spatial dimensions. For example, sound waves are governed by the wave equation in three dimensions,

$$\frac{\partial^2 p}{\partial t^2} = c^2 \left[\frac{\partial^2 p}{\partial x^2} + \frac{\partial^2 p}{\partial y^2} + \frac{\partial^2 p}{\partial z^2} \right], \qquad (21.14)$$

where p is the air pressure. If we build a numerical code with J subintervals in each of the three-spatial directions, the complete array of pressures would have $(N+1)(J+1)^3$ elements. With $N = 10000$ and $J = 500$ (not unreasonable numbers), this is more than 1.2×10^{12} elements.

21.3 Numerical instability

Exercise 21.3a

Change the timestep in your code to $\Delta t = 1.2\Delta x/c$. (Use $J = 100$. Keep the remaining parameters and the initial conditions unchanged.) Plot *separate* graphs of $y(x)$ at times $25\Delta t$ and $40\Delta t$.

As a general rule, a finite difference scheme for PDEs will become *numerically unstable* if the timestep Δt is too large. When this happens the numerical results are meaningless—the discrete values y_j^n do not approximate any solution of the original continuum PDE.

For the wave equation, $\Delta x/c$ is the time it takes a wave to travel from one spatial node to the next. The timestep is usually defined as a multiple of $\Delta x/c$, such as

$$\Delta t = \eta \Delta x/c. \tag{21.15}$$

Many numerical PDE methods will yield physically meaningful results only if η is small, less than some maximum value $\eta_{\max}$. The value of $\eta_{\max}$ depends on the details of the PDE and the details of the numerical algorithm, but typically $\eta_{\max}$ is of order unity. The restriction $\Delta t \leq \eta_{\max}\Delta x/c$ is called the *Courant* (or *Courant–Friedrichs–Lewy*) condition.

Exercise 21.3b

Run numerical experiments to find $\eta_{\max}$ (approximately) for the CTCS algorithm.

For some numerical algorithms, not considered here, $\eta_{\max}$ is infinite. Such a scheme is *unconditionally stable.* Although the scheme might be stable for arbitrarily large Δt, it is usually not accurate for large Δt. Some numerical algorithms are *unconditionally unstable*, that is, they are unstable for any value of Δt.

Let's take a close look at the instability in the CTCS method. First, we need to recognize that the wave equation (21.5) is a *linear* partial differential equation. If $y(t,x)$ and $\tilde{y}(t,x)$ are two solutions of the wave equation, then $y(t,x) + \epsilon\tilde{y}(t,x)$ is also a solution. Here, ϵ is a constant parameter. The same property holds for the discrete equation (21.9): if both y^n_j and $\tilde{y}^n_j$ satisfy the discrete wave equation, then $y^n_j + \epsilon\tilde{y}^n_j$ is also a solution.

Let y^n_j denote the physical solution of the discrete wave equation that we hope to find from our numerical code. An instability arises when the discrete wave equation also has an unphysical solution $\tilde{y}^n_j$ that grows without bound. The unphysical solution can become superimposed on top of the physical solution, so what the code actually finds is $y^n_j + \epsilon\tilde{y}^n_j$. Even if ϵ is very small, the amplitude of the unphysical term $\epsilon\tilde{y}^n_j$ can grow without bound and eventually overtake the physical solution.

The physical part of the solution, y^n_j, is determined by our choice of initial conditions. The unphysical part $\epsilon\tilde{y}^n_j$ creeps into the simulation due to machine roundoff errors. If the amplitude for the physical solution is of order unity (that is, typical values for y^n_j are around 1, to within a factor of ten or so) then the *initial* amplitude of the unphysical part $\epsilon\tilde{y}^n_j$ will be the size of machine roundoff error, roughly 10^{-16}. If we scale the unphysical solution $\tilde{y}^n_j$ so that its initial amplitude is of order unity, then $\epsilon \approx 10^{-16}$.

Figure 21.3 shows the results of a simulation of the wave equation using the CTCS method, with $\Delta t = 1.2\Delta x/c$. This violates the Courant condition. The initial data consist of a Gaussian pulse (21.11), initially at rest, with $A = 1$ and $\sigma = 1$. The pulse quickly splits into left-moving and right-moving pieces, each with amplitude 0.5. Superimposed on the physical solution (smooth, double pulse) is an unphysical part (the jagged spikes). The unphysical part of the solution shown in Fig. 21.3 grows very rapidly. At the time shown, $t = 0.072$, the amplitude of the spikes is ≈ 0.6. By the time $t = 0.1$, the spikes have grown to $\approx 10^6$. Even though the unphysical part of the solution begins with a small amplitude dictated by machine roundoff errors, it grows so quickly that the simulation is spoiled after only 30 timesteps.

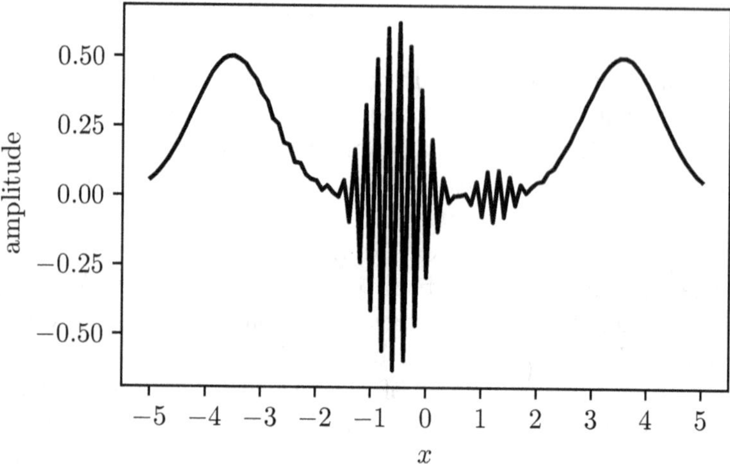

Fig. 21.3. Simulation of a Gaussian wave pulse using the CTCS method with Δt beyond the Courant limit. The figure shows the wave at time $t = 0.072$, after $n = 30$ timesteps.

Exercise 21.3c

Reproduce the simulation from Fig. 21.3 by setting (in SI units) $A = 1$, $\sigma = 1$, $c = 50$, $x_a = -10$, $x_b = 10$, and $J = 200$. Set the timestep to $\Delta t = 1.2\Delta x/c$. Use the NumPy function `np.max(y)` to determine the maximum amplitude at the end of each timestep. Plot the natural log of the maximum amplitude versus the timestep number n for $0 \leq n \leq 50$.

For $0 \leq n < 30$, the simulation is dominated by the physical solution so the maximum amplitude is of order unity. For $n > 30$, the simulation is dominated by the unphysical solution and the maximum amplitude grows exponentially.

Let $\epsilon\tilde{y}_{\max}(n)$ denote the maximum amplitude of the unphysical part of the solution, as a function of the timestep n. Since the unphysical solution grows exponentially, we have $\epsilon\tilde{y}_{\max}(n) = Ke^{sn}$ for some constants K and s. If the unphysical solution is "seeded" by machine roundoff errors in the early timesteps, we would expect that for very small n, $\epsilon\tilde{y}_{\max}(n)$ should be smaller than the physical signal by a factor of roughly 10^{-16}.

Exercise 21.3d

Determine the constants K and s in the formula $\epsilon\tilde{y}_{\max}(n) = Ke^{sn}$, using your data for $n \geq 32$. Use this result to compute $\epsilon\tilde{y}_{\max}(0)$ and $\epsilon\tilde{y}_{\max}(1)$. Are these on the order of machine error?

This calculation should help convince you that the amplitude of the unphysical part of the solution can grow from the size of machine error to order unity in just 30 or so timesteps. Instabilities don't always grow this quickly. Sometimes an unstable solution can take 1000's or more timesteps to overtake the physical solution and spoil the simulation.

21.4 An unphysical solution

When the Courant condition $\Delta t \leq \eta_{\max}\Delta x/c$ is violated, the discrete wave equation (21.9) admits many unphysical solutions. One such solution is fairly simple to write down:

$$\tilde{y}_j^n = \mathrm{Re}\Big[\big(1 - 2\eta^2 - 2\eta\sqrt{\eta^2 - 1}\big)^n\Big](-1)^j. \tag{21.16}$$

The symbol Re stands for the real part of the expression that follows. Recall from Eq. (21.15) that $\eta = c\Delta t/\Delta x$.

There are two cases to consider. If $\eta \geq 1$, then the square root term $\sqrt{\eta^2 - 1}$ is real. The entire expression in square brackets is real and the Re symbol can be dropped. On the other hand, if $\eta < 1$, the square root is imaginary: $\sqrt{\eta^2 - 1} = i\sqrt{1 - \eta^2}$. In this case, Re instructs us to take the real part of the complex expression in square brackets.

Exercise 21.4a

Verify that Eq. (21.16) is a solution of the discrete wave equation (21.9). Hint: it suffices to show that

$$\tilde{Y}_j^n = \left(1 - 2\eta^2 - 2\eta\sqrt{\eta^2 - 1}\right)^n (-1)^j$$

is a solution for any η. Since the wave equation is linear with real coefficients, the complex conjugate $(\tilde{Y}_j^n)^*$ must also be a solution. It follows that the real part $\tilde{y}_j^n = \mathrm{Re}\,\tilde{Y}_j^n = [\tilde{Y}_j^n + (\tilde{Y}_j^n)^*]/2$ is a solution.

The factor $(-1)^j$ in Eq. (21.16) is a clear indication that the solution is unphysical. The amplitude $\tilde{y}_j^n$ switches from positive to negative from one spatial node to the next, creating a "zig-zag" appearance in a plot of y versus x. This behavior depends crucially on our choices for the location and spacing of the spatial nodes x_j. Of course, a physical solution can have closely spaced peaks and valleys as well, but the distance between peaks and valleys should not be determined by the computer programmer's choice of node spacing.

Exercise 21.4b

Set the initial data in a CTCS code to match the unphysical solution:

$$\tilde{y}_j^0 = (-1)^j,$$

$$\tilde{y}_j^1 = \begin{cases} \left[1 - 2\eta^2\right](-1)^j, & \text{if } 0 < \eta \le 1, \\ \left[1 - 2\eta^2 - 2\eta\sqrt{\eta^2 - 1}\right](-1)^j, & \text{if } \eta > 1. \end{cases}$$

Plot $\tilde{y}_j^n$ versus j for various values of n, for both cases $0 < \eta \le 1$ and $\eta > 1$. Also plot $\tilde{y}_j^n$ versus n with a fixed value of j, for both $0 < \eta \le 1$ and $\eta > 1$.

Chapter 22

Partial Differential Equations II

22.1 First-order time derivatives

The wave equation (21.5) for a vibrating string contains second-order time derivatives. It can be rewritten as a system of equations with first-order time derivatives by introducing a new variable $v = \dot{y}$, which is the transverse velocity of the string. The resulting system of PDEs

$$\dot{y} = v, \tag{22.1a}$$

$$\dot{v} = c^2 y'' \tag{22.1b}$$

is equivalent to the single equation $\ddot{y} = c^2 y''$.

Using the two-point forward difference formula (16.4), the first time derivative of y is approximated as

$$\dot{y}(t^n, x_j) \approx \frac{y_j^{n+1} - y_j^n}{\Delta t}. \tag{22.2}$$

An analogous approximation is made for $\dot{v}$. With these results and the three-point centered difference approximation for y'', we have

$$y_j^{n+1} = y_j^n + \Delta t \, v_j^n, \tag{22.3a}$$

$$v_j^{n+1} = v_j^n + \Delta t \, c^2 \left(\frac{y_{j+1}^n - 2y_j^n + y_{j-1}^n}{\Delta x^2} \right). \tag{22.3b}$$

This is called the *forward-time, centered-space* (FTCS) discretization of the system (22.1).

The codes for the FTCS and CTCS algorithms are very similar. Here is an outline for an FTCS code:

- Import NumPy (as `np`) and other libraries.
- Choose values for the parameters.
- Set up the numerical grid as in the CTCS code.
- Define arrays:

```
y = np.zeros(J+1)       # y at time t^n
v = np.zeros(J+1)       # v at time t^n
yp = np.zeros(J+1)      # yp = "y plus" = y at time t^(n+1)
vp = np.zeros(J+1)      # vp = "v plus" = v at time t^(n+1)
```

- Choose initial data:

```
y = some function of x
v = some function of x
```

- Evolve the system forward in time using the FTCS scheme (22.3):

```
for n in range(1,N+1):
   for j in range(1,J):
      yp[j] = y[j] + dt*v[j]
      vp[j] = v[j] + (dt*c**2/dx**2)\
         *(y[j+1] - 2.0*y[j] + y[j-1])
   yp[0] = y[0]
   yp[J] = y[J]
   y = np.copy(yp)
   v = np.copy(vp)
```

  The endpoint values `v[0]` and `v[J]` are never used, so they do not need to be updated.
- Add code as necessary to plot graphs and print (or save) results.

Unfortunately, the FTCS algorithm is unconditionally unstable.

Exercise 22.1

Write a code for the wave equation using the FTCS scheme (22.3). As in the previous chapter, use mass per unit length $\mu = 0.01$ and tension $T = 25.0$ (in SI units), so the wave speed is $c = 50.0$. Set the boundaries at $x_a = -5.0$ and $x_b = 5.0$. For initial conditions, use the Gaussian pulse at rest. That is, y_j^0 is given by Eq. (21.12) and $v_j^0 = 0$. Experiment with different timestep values. Have your code display a graph of $y(x)$ showing the unphysical results.

22.2 Lax–Friedrich method

The FTCS algorithm can be modified to make it stable. Replace y_j^n on the right-hand side of Eq. (22.3a) with the average $(y_{j+1}^n + y_{j-1}^n)/2$. The resulting method,

$$y_j^{n+1} = \frac{1}{2}(y_{j+1}^n + y_{j-1}^n) + \Delta t\, v_j^n, \tag{22.4a}$$

$$v_j^{n+1} = v_j^n + \Delta t\, c^2 \left(\frac{y_{j+1}^n - 2y_j^n + y_{j-1}^n}{\Delta x^2} \right), \tag{22.4b}$$

is stable, provided the timestep satisfies a Courant condition. We will call this the Lax–Friedrichs (LF) method.[1]

Exercise 22.2a

Solve the wave equation based on the LF algorithm. Use the same initial data as in the previous exercise. Choose a relatively small timestep so that the simulation is stable. Plot y versus x at various times.

---

[1]The term "Lax–Friedrichs method" usually refers to a modification of the FTCS algorithm for the advection equation $\dot{y} = cy'$. Our modification for the wave equation is similar.

Exercise 22.2b

Experiment with your LF code to obtain an estimate of the Courant condition $\Delta t \leq \eta_{\rm max}\Delta x/c$. (Hint: The answer is $\eta_{\rm max} = 1/\sqrt{2}$.)

22.3 Energy of a vibrating string

In the absence of dissipating effects like air resistance, the energy of a vibrating string is

$$E = \frac{1}{2}\int_{x_a}^{x_b} dx\, \left[\mu\, v^2 + T(y')^2\right], \tag{22.5}$$

where $v = \dot{y}$. Recall that μ is the mass per unit length and T is the tension. The two terms in E are the kinetic and elastic potential energies of the string. This might not be entirely obvious, but it is easy to verify that E is conserved. Start by differentiating E with respect to time, and bring the time derivative under the integral over x:

$$\begin{aligned} \frac{dE}{dt} &= \frac{1}{2}\int_{x_a}^{x_b} dx \frac{\partial}{\partial t}\left[\mu\, v^2 + T(y')^2\right] \\ &= \int_{x_a}^{x_b} dx\, \left[\mu v\dot{v} + Ty'\dot{y}'\right]. \end{aligned} \tag{22.6}$$

Since $\dot{v} = c^2 y''$, this simplifies to

$$\frac{dE}{dt} = \int_{x_a}^{x_b} dx\, \left[\mu c^2 v y'' + Ty'v'\right]. \tag{22.7}$$

where $v' = \dot{y}'$. Now, because $c^2 = T/\mu$, we can remove a common factor of T from the integral and write

$$\frac{dE}{dt} = T\int_{x_a}^{x_b} dx \frac{\partial}{\partial x}\left[vy'\right]. \tag{22.8}$$

This reduces to a boundary term,

$$\frac{dE}{dt} = T\left[vy'\right]\Big|_{x_a}^{x_b}. \tag{22.9}$$

Since y is fixed at the boundaries, the time derivative $v = \dot{y}$ must vanish at x_a and x_b. It follows that $dE/dt = 0$; the energy is constant in time.

Although E is a constant, the energy calculated from a numerical code will not remain constant due to truncation errors. Thus, we can use the energy to help monitor the accuracy of our code. To compute E, we first need to discretize the expression (22.5). The integral over x can be replaced by a simple midpoint rule approximation. The string velocity at the midpoint of the jth subinterval is (to second-order accuracy) just the average of the velocities at the adjacent nodes: $(v_j + v_{j-1})/2$. The derivative y' at the midpoint, half-way between nodes $j-1$ and j, is given to second-order accuracy by the three-point centered difference formula $(y_j - y_{j-1})/\Delta x$. This yields the discrete expression

$$E = \frac{1}{2}\sum_{j=1}^{J} \Delta x \left[\mu \left(\frac{v_j + v_{j-1}}{2} \right)^2 + T \left(\frac{y_j - y_{j-1}}{\Delta x} \right)^2 \right], \tag{22.10}$$

for the energy of the string.

Exercise 22.3

Compute the energy at the end of each timestep in your LF code. Plot the energy as a function of time.

22.4 Method of lines

The LF method uses the two-point forward difference formula to approximate time derivatives. This difference formula is first-order accurate, so it generally limits the LF method to first-order accuracy. Assuming the timestep obeys the Courant condition, the errors for the LF method are proportional to the first power of Δt. Since Δt is proportional to Δx, which in turn is proportional to $1/J$, we see that the errors are proportional to $1/J$. We can verify this using a three-point convergence test as discussed in Secs. 17.5 and 19.3.

Let $E_{(J)}$ denote the energy of the string computed numerically at some fixed final time with resolution J. Since the errors are first order,

$$E_{(J)} = E_{(\text{exact})} + K/J \tag{22.11}$$

for some constant K. Here, $E_{(\text{exact})}$ is the (unknown) exact answer. This relation holds for any value of J, which implies

$$\frac{E_{(J)} - E_{(2J)}}{E_{(2J)} - E_{(4J)}} = 2. \tag{22.12}$$

The result is approximate because Eq. (22.11) only holds in the limit of high resolution. Thus, if we run the code at three successive resolutions and compute the left-hand side of Eq. (22.12), the result should approach 2 as J is increased.

Exercise 22.4a

Carry out the three-point convergence test with $x_a = -5$, $x_b = 5$, and $c = 50$. Set the timestep to $\Delta t = 0.25\Delta x/c$ and use the time interval $0 \leq t \leq 0.25$. For initial conditions, use the Gaussian pulse at rest. Determine $E_{(J)}$ for resolution $J = 100$, 200, 400, etc. Does the left-hand side of Eq. (22.12) approach 2 in the limit of high resolution?

The LF method is not very accurate unless the resolution is very high. There are better methods for solving PDEs. The challenge is to find a higher-order algorithm that is numerically stable.

One general approach is based on the *method of lines*. Here's the idea. Start by discretizing the continuum differential equations (22.1) in space. Define $y_j(t) = y(t, x_j)$ and $v_j(t) = v(t, x_j)$, where, as usual, x_j are the spatial nodes (with $j = 0, \ldots, J$). Using the second-order finite difference approximation for y'', the PDEs become

$$\dot{y}_j = v_j, \tag{22.13a}$$

$$\dot{v}_j = c^2(y_{j+1} - 2y_j + y_{j-1})/\Delta x^2. \tag{22.13b}$$

This is a set of coupled *ordinary* differential equations (ODEs), with independent variable t and dependent variables $y_0, y_1, \ldots, y_J$ and $v_0, v_1, \ldots, v_J$. We can solve them using the ODE methods from earlier chapters.

The method of lines approach does not guarantee that the resulting algorithm is stable. If we use Euler's method to solve Eqs. (22.13), the result is equivalent to the FTSC method. As we saw in Sec. 22.1, this is unconditionally unstable. We could apply second-order Runge–Kutta to solve Eqs. (22.13), but that turns out to be unconditionally unstable as well.

Fourth-order Runge Kutta is stable, provided the Courant condition $\Delta t \leq \sqrt{2}\Delta x/c$ is satisfied. We can apply the RK4 algorithm (18.12) to the system (22.13) by modifying an FTCS code. Simply replace the section of code inside the loop over n with the following. First, compute the variables denoted u_a and u_b in Eqs. (18.12):

```
for j in range(1,J):
    ya[j] = y[j] + 0.5*dt*v[j]
    va[j] = v[j] + 0.5*dt*(c**2/dx**2)*(y[j+1]-2.0*y[j]+y[j-1])
ya[0] = y[0]
ya[J] = y[J]
for j in range(1,J):
    yb[j] = y[j] + 0.5*dt*va[j]
    vb[j] = v[j] + 0.5*dt*(c**2/dx**2)*(ya[j+1]-2.0*ya[j]+ya[j-1])
yb[0] = y[0]
yb[J] = y[J]
```

Continue with the analogous calculations for u_c and u_d, then compute the string amplitude and velocity at time t^{n+1}:

```
for j in range(0,J+1):
    yp[j] = (1/3)*(ya[j] + 2*yb[j] + yc[j] + yd[j]/2) - y[j]/2
    vp[j] = (1/3)*(va[j] + 2*vb[j] + vc[j] + vd[j]/2) - v[j]/2
```

Finally, copy the arrays `yp` and `vp` into `y` and `v` in preparation for the next timestep:

```
y = np.copy(yp)
v = np.copy(vp)
```

Exercise 22.4b

Solve the wave equation using the method of lines with fourth-order Runge–Kutta. Plot the energy as a function of time. How do these results compare to the results you obtained with the LF method?

As a time-evolution scheme, RK4 is fourth-order accurate. But the discretization in space using the three-point finite difference stencil is only second-order accurate. Also keep in mind that the discretization (22.10) of the energy has second-order errors.

Exercise 22.4c

Use Simpson's rule to compute the energy integral (22.5) when the initial data consist of a Gaussian pulse at rest. Use a sufficiently large number of subintervals to obtain the answer accurate to 8 or more significant figures. Treat this result as the "exact" answer and compute the error in the energy for the RK4 algorithm at various resolutions. Estimate the time average of the error at each resolution. What is the order of convergence?

22.5 Other initial conditions

The initial conditions can be chosen arbitrarily as long as the boundary conditions are met. Since the string is fixed at the endpoints, the initial conditions should obey $v(0, x_a) = 0$ and $v(0, x_b) = 0$.

Exercise 22.5a

Use the method of lines with RK4 to solve the wave equation with initial conditions

$$\begin{aligned} y(0,x) &= 2\sin(2\pi(x - x_a)/(x_b - x_a)) \\ &\quad + \sin(5\pi(x - x_a)/(x_b - x_a)), \\ v(0,x) &= 20\sin(3\pi(x - x_a)/(x_b - x_a)). \end{aligned}$$

Experiment with other initial conditions.

The general solution to the wave equation is a sum of left-moving and right-moving waves, as shown in Eq. (21.6). Consider a right-moving Gaussian pulse

$$y(t,x) = Ae^{-(x-ct)^2/\sigma^2}. \tag{22.14}$$

The initial data for this solution is

$$y(0,x) = Ae^{-x^2/\sigma^2}, \tag{22.15a}$$

$$v(0,x) = (2Acx/\sigma^2)e^{-x^2/\sigma^2}, \tag{22.15b}$$

where $v = \dot{y}$. These conditions do not satisfy $v(0,x_a) = 0 = v(0,x_b)$. However, if σ is not too small, the string velocity will be very close to 0 at the endpoints.

Exercise 22.5b

Solve the wave equation with the initial data (22.15) using the method of lines with RK4. Plot y versus x at various times and describe the results. What happens if you make a mistake in the formula for $v(0,x)$? For example, you omit the factor of 2? Or replace σ^2 with 3σ in the exponent?

22.6 Other boundary conditions

Up to this point we have considered the string amplitude to be fixed at the endpoints x_a and x_b. Conditions of this type, with the dependent variable fixed at the boundaries, are called *Dirichlet boundary conditions.*[2]

An alternative is *Neumann boundary conditions* in which the spatial derivative of y is fixed at the boundaries. We can implement Neumann boundary conditions by tying each end of the string to a massless ring, and letting each ring slide (without friction) on a post. Consider the ring at the right end, x_b, shown in Fig. 22.1. Because the ring is massless, the net force on the ring must vanish. Otherwise the ring would have an infinite acceleration. Now, the post cannot exert a vertical force on the ring because the ring slides without friction. The only other force on the ring is from tension in the string, so this force must be horizontal. In other words, the slope of the string at x_b must vanish.

[2]In general, with Dirichlet boundary conditions, the endpoint values $y(t,x_a)$ and $y(t,x_b)$ may be time dependent.

Fig. 22.1. Neumann boundary conditions. The string is attached to a massless ring that slides without friction on a vertical post. The slope of the string vanishes at the point of attachment.

Of course the same reasoning holds at the left end. The slope of the string must vanish at both ends, so that $y'(t, x_a) = y'(t, x_b) = 0$. These are Neumann boundary conditions.

You can switch from Dirichlet to Neumann boundary conditions in your RK4 code by replacing `ya[0] = y[0]` with `ya[0] = ya[1]` and replacing `ya[J] = y[J]` with `ya[J] = ya[J-1]` at the end of the first sub-step. Make analogous changes at the end of the second, third and fourth sub-steps.

Exercise 22.6

Create a version of your RK4 code with Neumann boundary conditions. Experiment with different initial conditions. What happens when a pulse hits the boundary? How does this differ from Dirichlet boundary conditions?

22.7 Other PDEs

We can easily modify these techniques to solve other systems of PDEs. Unfortunately, there is no guarantee that the algorithm will be stable.

Consider the damped wave equation $\ddot{y} + k\dot{y} = c^2 y''$ with damping constant k. This PDE can be written as the first-order system

$$\dot{y} = v, \tag{22.16a}$$

$$\dot{v} = c^2 y'' - kv. \tag{22.16b}$$

Exercise 22.7a

Numerically solve Eqs. (22.16) using the method of lines with RK4. Use Dirichlet or Neumann boundary conditions and any initial conditions. Plot a graph of y versus x at various times using $c = 50$ and $k = 20$ (SI units).

The *heat equation* describes the temperature T in a body as a function of time and spatial location. Consider a cylindrical rod that extends along the x-axis from x_a to x_b. The sides of the rod are insulated. We will assume that the temperature $T(t, x)$ depends only on t and x—that is, the temperature is constant along each cross-section of the rod. The heat equation for the rod is

$$\frac{\partial T}{\partial t} = \alpha \frac{\partial^2 T}{\partial x^2}, \tag{22.17}$$

where α is the *thermal diffusivity* of the material.

We can impose various boundary conditions at the ends of the rod. For example, we can attach a large body at temperature T_a to the left end. This imposes the Dirichlet boundary condition $T(t, x_a) = T_a$. We can insulate the right end. This imposes the Neumann boundary condition $\partial T(t, x_b)/\partial x = 0$.

The heat equation can be solved numerically using FTCS (forward differencing in time, centered differencing in space). For stability, the Courant condition $\Delta t \leq \Delta x^2/(2\alpha)$ must be satisfied.

Exercise 22.7b

Use the FTCS method to solve the heat equation for an aluminum rod of length $x_b - x_a = 1.0\,\text{m}$ and thermal diffusivity $\alpha = 9.7 \times 10^{-5}\text{m}^2/\text{s}$. Impose the Dirichlet boundary condition at the left end with $T_a = 350\,\text{K}$, and impose the Neumann boundary condition at the right end. For initial conditions, choose

$$T(0, x) = 270 + 80e^{-100(x-x_a)^2}.$$

Plot $T(t, x)$ versus x for various times t. How long does it take for the temperature at the right end of the rod to reach 280 K? 300 K? 320 K?

22.8 Animation

Animation of a time-series simulation is fun, and can lead to helpful insights. To animate your code:

- Remove old plot commands.
- Import libraries:

```
# %matplotlib ipympl      # uncomment if needed
import numpy as np
import matplotlib.pyplot as plt
import matplotlib.animation as anim
```

- Define parameters, constants, arrays, and initial data as before. Then add:

```
plt.close()
fig, ax = plt.subplots()
curve = ax.plot(x,y)[0]
```

In general, the function `ax.plot()` returns a list of curves to plot. In this case the list has just one element. The command `ax.plot(x,y)[0]` extracts this element (with index 0) from the list.

- At the core of your previous codes you have a loop that steps through time. For example,

```
for n in range(1,N+1):
   compute y and v at time t^{n+1} from y and v at time t^n
```

Replace this loop with the following function definition:

```
def update(num):
   global y
   global v
   compute y and v at time t^{n+1} from y and v at time t^n
   curve.set_data(x,y)
   return
```

- Add this to the end of your code:

```
myanim = animation.FuncAnimation(fig, update, N,
      interval=50, repeat=False)
# plt.show()      # uncomment if needed
```

The arguments of `FuncAnimation()` are:

- `fig` (the name of the figure)
- `update` (the name of the function that updates the figure)
- `N` (the number of timesteps)
- `interval=50` (sets the time interval between each frame to 50 ms)
- `repeat=False` (if set to `True`, the animation will repeat)

- To save the animation as a gif, add the statement

```
myanim.save('my_movie.gif')
```

to the end of your code.

At the time of writing, this prescription works well in most integrated development environments. (You might need to uncomment the `plt.show()` statement.) Animation is not quite as robust in Jupyter notebooks, due to the extra layer of interaction between the Python interpreter and the web browser. If animation is not working in a Jupyter notebook, try uncommenting the line `%matplotlib ipympl` at the beginning of the code. This command tells Python to use `ipympl` to translate the Matplotlib graphics commands into javascript commands that can be rendered in a web browser. If your code complains that `ipympl` is not installed, you will need to install it.

Exercise 22.8a

Create a version of your wave equation code with animation. Use the initial conditions from Exercise 22.5a with Dirichlet boundary conditions.

Exercise 22.8b

Animate the waves on a string with initial conditions

$$\begin{aligned} y(0,x) &= \cos(5\pi(x-x_a)/(x_b-x_a)) \\ &\quad + \cos(3\pi(x-x_a)/(x_b-x_a)), \\ v(0,x) &= 25\cos(2\pi(x-x_a)/(x_b-x_a)), \end{aligned}$$

and Neumann boundary conditions.

22.9 First-order form of the wave equation

The wave equation can be reduced to a system of PDEs with first-order derivatives in both time and space. Before describing the reduction, let's review what we know about the wave equation in its original form, $\ddot{y} = c^2 y''$, with second-order time and space derivatives. To be definite we will consider time-independent Dirichlet boundary conditions. Thus, the amplitude of the string at each endpoint is a prescribed constant:

$$y(t, x_a) = \text{const}, \tag{22.18a}$$

$$y(t, x_b) = \text{const}. \tag{22.18b}$$

The initial data consist of the initial amplitude and velocity:

$$y(0, x) = \text{function of } x, \tag{22.19a}$$

$$\dot{y}(0, x) = \text{function of } x. \tag{22.19b}$$

These functions can be chosen freely as long as they satisfy the boundary conditions. That is, the initial data function $y(0, x)$ evaluated at x_a must agree with the constant (22.18a). Likewise, $y(0, x)$ evaluated at x_b must agree with the constant (22.18b). Also observe that the time derivatives of Eqs. (22.18) imply $\dot{y}(t, x_a) = 0$ and $\dot{y}(t, x_b) = 0$. The initial data function $\dot{y}(0, x)$ must satisfy these relations as well.

The reduction of the wave equation to first-order time and space derivatives begins with the definitions

$$v = \dot{y}, \tag{22.20a}$$

$$w = cy'. \tag{22.20b}$$

In terms of these new variables, the wave equation becomes

$$\dot{v} = cw'. \tag{22.21}$$

Together, these three equations are equivalent to the original wave equation.

Our next task is to express the boundary conditions and initial conditions in terms of the new variables y, v, and w. Of course the initial conditions (22.19a) are unchanged, but $\dot{y}$ now has the new name v. Thus, the initial conditions become

$$y(0,x) = \text{function of } x, \tag{22.22a}$$

$$v(0,x) = \text{function of } x. \tag{22.22b}$$

Since $y(0,x)$ is a chosen function of x, we can compute its spatial derivative $y'(0,x)$. Therefore, by the definition (22.20b), w must satisfy

$$w(0,x) = cy'(0,x). \tag{22.23}$$

Thus, the initial value for w is also determined by the initial data.

The boundary conditions tell us that the endpoints of the string are fixed; that is, Eqs. (22.18) hold. Since the endpoint values are constants in time, we also find

$$v(t,x_a) = 0, \tag{22.24a}$$

$$v(t,x_b) = 0. \tag{22.24b}$$

In turn, the values of v at the endpoints are constants in time (equal to zero), so the wave equation (22.21) implies

$$w'(t,x_a) = 0, \tag{22.25a}$$

$$w'(t,x_b) = 0. \tag{22.25b}$$

These are Neumann boundary conditions for the variable w.

Observe that Eq. (22.20b) doesn't contain any time derivatives. How do we handle this in a numerical code? Since this equation must hold for all time, we can differentiate with respect to t to obtain $\dot{w} = c\dot{y}'$. Ordinarily, we would not be allowed to replace $w = cy'$ with $\dot{w} = c\dot{y}'$, because these equations are not equivalent. In fact, $\dot{w} = c\dot{y}'$ implies $w = cy' + f(x)$, where $f(x)$ is some function of x but is constant in t. However, the initial condition (22.23) tells us that $f(x)$ vanishes. We can replace $w = cy'$ with $\dot{w} = c\dot{y}'$, provided we also impose the initial condition (22.23). Finally, note that, since $\dot{y} = v$, we can write this new equation as $\dot{w} = cv'$.

To summarize, the wave equation can be written as a system of PDE's with first-order time and space derivatives as

$$\dot{y} = v, \tag{22.26a}$$

$$\dot{v} = cw', \tag{22.26b}$$

$$\dot{w} = cv'. \tag{22.26c}$$

The initial conditions are listed in Eqs. (22.22) and (22.23). The boundary conditions are listed in Eqs. (22.18), (22.24), and (22.25).

Exercise 22.9a

Solve the first-order system (22.26) using the FTCS method:

$$y_j^{n+1} = y_j^n + \Delta t v_j^n,$$
$$v_j^{n+1} = v_j^n + c\Delta t(w_{j+1}^n - w_{j-1}^n)/(2\Delta x),$$
$$w_j^{n+1} = w_j^n + c\Delta t(v_{j+1}^n - v_{j-1}^n)/(2\Delta x).$$

This method is unconditionally unstable, but you should get reasonable results for short run times.

Exercise 22.9b

Apply the Lax–Friedrich modification to your previous code: Replace v_j^n and w_j^n with $(v_{j+1}^n + v_{j-1}^n)/2$ and $(w_{j+1}^n + w_{j-1}^n)/2$ in the second and third equations. (You do not need to modify the first equation. Can you guess why?) This method is stable as long as the Courant condition $\Delta t \leq \Delta x/c$ is satisfied. Graph y versus x at various times t, and compare results to your RK4 code.

Chapter 23

Fourier Analysis

Fourier analysis is a powerful tool. We often deal with data that consist of a time-dependent signal, such as an audio signal or an electromagnetic signal. Fourier analysis tells us which frequencies make up the signal.

23.1 Fourier series

Any periodic signal can be described as an infinite sum of sine and cosine functions. Consider a signal $S(t)$ with period T; that is, $S(t+T) = S(t)$. We will assume that the signal is real valued, as opposed to complex. Define

$$\omega_n = 2\pi n/T, \tag{23.1}$$

where $n = 0, 1, \ldots, \infty$. The signal $S(t)$ can be written as a *Fourier series*

$$S(t) = \frac{1}{2}C_0 + \sum_{n=1}^{\infty} \bigl[C_n \cos(\omega_n t) + D_n \sin(\omega_n t)\bigr]. \tag{23.2}$$

The constant coefficients C_n and D_n are the *Fourier amplitudes.* Each term in the Fourier series is referred to as a *mode*. The signal $S(t)$ consists of a constant mode and oscillatory modes. The constant mode, or "zero mode," consists of the term $C_0/2$ with angular frequency $\omega_0 = 0$.[1] The oscillatory modes are the cosine and sine terms with angular frequencies ω_1, ω_2, etc.

Exercise 23.1a

Compute the signal $S(t)$ with period T and Fourier amplitudes C_n and D_n given by

```
T = 2.5
C = np.array([1,3,2,5,-2,7])
D = np.array([0,2,-1,3,4,-5])
```

Graph $S(t)$ versus t. Note that the array element `D[0]` is irrelevant, since D_0 does not appear in the series (23.2).

Given a signal $S(t)$, how do we determine the amplitudes C_n and D_n? Standard results from Fourier analysis yield

$$C_n = \frac{2}{T}\int_0^T dt\, S(t)\, \cos(\omega_n t), \quad n = 0, 1, \ldots, \infty, \tag{23.3a}$$

$$D_n = \frac{2}{T}\int_0^T dt\, S(t)\, \sin(\omega_n t), \quad n = 1, 2, \ldots, \infty. \tag{23.3b}$$

If the Fourier series (23.2) for $S(t)$ is inserted into the right-hand sides of Eqs. (23.3), the right-hand sides simplify to C_n and D_n.

Exercise 23.1b

Compute the Fourier amplitudes (23.3) with $n = 0, \ldots, 10$ using the signal $S(t)$ obtained in the previous exercise. Give these amplitudes new names, for example, c_n and d_n. You can use the `scipy.integrate` function `quad()` to compute the integrals. Compare your results for c_n and d_n to the original amplitudes C_n and D_n.

[1] Angular frequency ω (radians per second) is related to frequency f (cycles per second) by $\omega = 2\pi f$. We use angular frequency through this chapter, although sometimes ω will be loosely referred to as the frequency.

The converse holds. If the definitions (23.3) for C_n and D_n are inserted into the right-hand side of the Fourier series (23.2), the right-hand side simplifies to $S(t)$.

Exercise 23.1c

Consider the signal

$$S(t) = \cos(1 + 2\sin(t + 1/2)),$$

which has period $T = 2\pi$. Plot $S(t)$ versus t and compute the Fourier amplitudes for $n = 0, \ldots, 7$.

For a general signal $S(t)$, there are an infinite number of Fourier amplitudes C_n and D_n. We can't compute all of them. Often, as in the example above, the amplitudes become vanishingly small as the angular frequency ω_n increases. In the previous exercise you were asked to compute the amplitudes through $n = 7$. How closely does the Fourier series (23.2) reproduce the signal $S(t)$ when the terms beyond $n = 7$ are omitted?

Exercise 23.1d

Use the C's and D's from the previous exercise to compute

$$\tilde{S}(t) = \frac{1}{2}C_0 + \sum_{n=1}^{7}\left[C_n \cos(\omega_n t) + D_n \sin(\omega_n t)\right].$$

Graph $\tilde{S}(t)$ and the original signal $S(t)$. Is the Fourier series well represented by the terms $n \leq 7$?

The average *power* for a real-valued, periodic signal $S(t)$ is defined by

$$P = \frac{1}{T}\int_0^T (S(t))^2\, dt. \tag{23.4}$$

If we insert the Fourier series into this expression and use the orthogonality properties of sines and cosines, we find

$$P = \frac{1}{4}C_0^2 + \frac{1}{2}\sum_{n=1}^{\infty}[C_n^2 + D_n^2]. \tag{23.5}$$

The power is a sum of contributions from each mode, $P = \sum_{n=0}^{\infty} P_n$, where

$$P_0 = \frac{1}{4}C_0^2, \tag{23.6a}$$

$$P_n = \frac{1}{2}[C_n^2 + D_n^2], \quad n = 1, \ldots, \infty. \tag{23.6b}$$

P_0 is the contribution from the zero mode. $P_1, P_2, \ldots$ are the contributions from the oscillatory modes. We refer to the P's as the *power spectrum*.

Exercise 23.1e

For the signal $S(t) = \cos(1 + 2\sin(t + 1/2))$ used in the previous exercises, compute the power spectrum for the first eight modes, $n = 0, \ldots, 7$.

23.2 Discrete Fourier transform

Often the time domain signal $S(t)$ is know only at discrete points in time. The signal might come from an experiment where the detectors and instruments record the signal at discrete times. The signal might come from a numerical simulation where the time evolution takes place in discrete steps.

Consider a real valued signal that is sampled at times $t = 0, \Delta t$, $2\Delta t$, etc., up to $(N-1)\Delta t$, where N is the total number of samples. For now, we will assume that N is odd. We can write the discrete times as

$$t_k = k\Delta t, \quad k = 0, \ldots, N-1, \tag{23.7}$$

and define the frequencies by

$$\omega_n = \frac{2\pi n}{N\Delta t}, \tag{23.8}$$

where n is a nonnegative integer.

Denote the signal at time t_k by S_k; that is, let $S_k = S(t_k)$ The discrete signal can be written as

$$S_k = \frac{1}{2}C_0 + \sum_{n=1}^{(N-1)/2} \left[C_n \cos(\omega_n t_k) + D_n \sin(\omega_n t_k)\right], \tag{23.9}$$

where the Fourier amplitudes are

$$C_n = \frac{2}{N}\sum_{k=0}^{N-1} S_k \cos(\omega_n t_k), \tag{23.10a}$$

$$D_n = \frac{2}{N}\sum_{k=0}^{N-1} S_k \sin(\omega_n t_k). \tag{23.10b}$$

Equations (23.10) define the *discrete Fourier transform* (DFT) of the signal S_k. Equation (23.9) is the inverse DFT.

Compare the definitions (23.8), (23.9), (23.10) to Eqs. (23.1), (23.2), (23.3) for the Fourier series. The formal correspondence

$$t \longrightarrow t_k,$$

$$T \longrightarrow N\Delta t,$$

$$\sum_{n=1}^{\infty} \longrightarrow \sum_{n=1}^{(N-1)/2},$$

$$\int_0^T dt \longrightarrow \sum_{k=0}^{N-1} \Delta t,$$

converts the formulas for the Fourier series into the formulas for the DFT.

The inverse DFT (23.9) "undoes" the discrete Fourier transform (23.10). If the amplitudes (23.10) are inserted into the right-hand side of the series (23.9), the result is the discrete signal S_k. Likewise, when the series (23.9) is inserted into the right-hand sides of Eqs. (23.10), the results are C_n and D_n.

Exercise 23.2a

The data from Table 23.1 are contained in the file *SignalData.txt* (see the Appendix). Compute the DFT amplitudes (23.10) for this data, for $n = 0$ through $n = (N-1)/2$. Use your results for C_n and D_n to compute the inverse DFT Eq. (23.9). Does this match the signal from Table 23.1?

Exercise 23.2b

For the data in Table 23.1 (*SignalData.txt*), plot the time-domain signal S_k versus t_k. Plot the Fourier amplitudes C_n and D_n versus ω_n for $n = 1, \ldots (N-1)/2$.

The formal correspondence between the Fourier series and the DFT can be used to obtain the average power:

$$P = \frac{1}{N} \sum_{k=0}^{N-1} (S_k)^2. \tag{23.11}$$

In terms of Fourier amplitudes,

$$P = \frac{1}{4} C_0^2 + \frac{1}{2} \sum_{n=1}^{(N-1)/2} \left[C_n^2 + D_n^2 \right]. \tag{23.12}$$

Table 23.1. A signal S_k sampled at $N = 11$ discrete times separated by $\Delta t = 0.1$.

k	0	1	2	3	4	5	6	7	8	9	10
t_k	0.0	0.1	0.2	0.3	0.4	0.5	0.6	0.7	0.8	0.9	1.0
S_k	−0.4	5.3	8.7	6.4	0.3	−2.6	−3.3	−5.4	5.2	2.4	1.6

This result can be derived directly from Eq. (23.11) using the inverse DFT (23.9) and the orthogonality properties of sines and cosines. We obtain

$$P_0 = \frac{1}{4}C_0^2, \tag{23.13a}$$

$$P_n = \frac{1}{2}\left[C_n^2 + D_n^2\right], \quad n = 1, \ldots, (N-1)/2, \tag{23.13b}$$

for the power spectrum of the discrete signal S_k.

Exercise 23.2c

Compute the power spectrum for the signal from Table 23.1 and *SignalData.txt*. Plot P_n as a function of angular frequency ω_n.

23.3 Complex form of the DFT

The discrete Fourier transform can be put into a more compact and elegant form by combining the real Fourier amplitudes C_n and D_n into a single complex amplitude

$$A_n = \frac{N}{2}(C_n - iD_n), \tag{23.14}$$

where $i = \sqrt{-1}$ is the imaginary unit.

Recall that D_0 does not appear in the inverse DFT (23.9). Consistent with the DFT (23.10), we take $D_0 = 0$. It follows that the zero-mode amplitude is real, $A_0 = (N/2)C_0$.

Using the results (23.10) for C_n and D_n, we find

$$A_n = \sum_{k=0}^{N-1} S_k\left[\cos(\omega_n t_k) - i\sin(\omega_n t_k)\right]. \tag{23.15}$$

The factor in square brackets can be simplified by using Euler's formula

$$e^{i\theta} = \cos\theta + i\sin\theta, \tag{23.16}$$

which holds for any θ. The complex amplitude reduces to

$$A_n = \sum_{k=0}^{N-1} S_k e^{-i\omega_n t_k}. \tag{23.17}$$

This is the DFT in complex form.

Recall that C_n and D_n are real, since the discrete signal S_k is (by assumption) real. Then the complex conjugate of A_n is

$$A_n^* = \frac{N}{2}(C_n + iD_n). \tag{23.18}$$

Combining this result with the definition (23.14), we find

$$C_n = \frac{1}{N}(A_n + A_n^*), \tag{23.19a}$$

$$D_n = \frac{i}{N}(A_n - A_n^*), \tag{23.19b}$$

and the inverse DFT (23.9) becomes

$$\begin{aligned} S_k &= \frac{1}{2N}(A_0 + A_0^*) \\ &\quad + \frac{1}{N}\sum_{n=1}^{(N-1)/2} \left[(A_n + A_n^*)\cos(\omega_n t_k) + i(A_n - A_n^*)\sin(\omega_n t_k)\right]. \end{aligned} \tag{23.20}$$

Since A_0 is real, the term $A_0 + A_0^*$ can be written as $2A_0$. Then

$$S_k = \frac{1}{N}A_0 + \frac{1}{N}\sum_{n=1}^{(N-1)/2}\left[A_n e^{i\omega_n t_k} + A_n^* e^{-i\omega_n t_k}\right], \tag{23.21}$$

where we have used Euler's formula once again.

Before continuing, we need to derive an important identity. Remember that N is odd. Equation (23.21) shows that the amplitudes A_n with $0 \leq n \leq (N-1)/2$ contain complete information about the signal—they can be used to reconstruct S_k. Nevertheless, the definitions (23.17) and (23.8) for the DFT and angular frequency can be extended to index values beyond $(N-1)/2$. With this in mind, rewrite Eq. (23.17) as

$$A_{N-n} = \sum_{k=1}^{N-1} S_k e^{-i\omega_{N-n} t_k}, \tag{23.22}$$

where $1 \leq n \leq (N-1)/2$. Since the frequencies and discrete times are $\omega_n = 2\pi n/(N\Delta t)$ and $t_k = k\Delta t$, it follows that $\omega_{N-n} t_k = 2\pi k - \omega_n t_k$. Because $e^{2\pi k} = 1$ for integer values of k, the exponential factor in Eq. (23.22) reduces to $\exp(-\omega_n t_k)$. We see that, since S_k is real, the right–hand side of Eq. (23.22) is simply A_n^*. This proves the identity

$$A_{N-n} = A_n^*. \tag{23.23}$$

The complex amplitudes with "large" indices in the range $(N+1)/2 \leq n \leq N-1$ are the complex conjugates of the amplitudes with "small" indices in the range $1 \leq n \leq (N-1)/2$.

Remember, in Python the imaginary unit is `j`. A complex number such as $3 + 7i$ is written as `3 + 7j`, with no asterisk between 7 and `j`. To create a NumPy array `A` with complex data type, use (for example)

```
A = np.zeros(N,dtype=complex)
```

Also note that the complex conjugate and absolute value of `z` are `np.conjugate(z)` and `np.absolute(z)`, respectively.

Exercise 23.3a

Compute the complex amplitudes A_n for $0 \leq n \leq N-1$ using the signal from Table 23.1 (*SignalData.txt*). Verify the identity (23.23).

With the identity (23.23) the inverse DFT (23.21) can be written as

$$S_k = \frac{1}{N} A_0 + \frac{1}{N} \sum_{n=1}^{(N-1)/2} A_n e^{i\omega_n t_k} + \frac{1}{N} \sum_{n=1}^{(N-1)/2} A_{N-n} e^{-i\omega_n t_k}. \tag{23.24}$$

Focus on the sum in the last term. Replace the summation index n with m, where $n = N - m$. As n ranges from 1 to $(N-1)/2$, the new index m ranges from $N-1$ to $(N+1)/2$. Thus, we have

$$\sum_{n=1}^{(N-1)/2} A_{N-n} e^{-i\omega_n t_k} = \sum_{m=N-1}^{(N+1)/2} A_m e^{-i\omega_{N-m} t_k}. \tag{23.25}$$

Again use the results $\omega_{N-m} t_k = 2\pi k - \omega_m t_k$, and $e^{2\pi k} = 1$ for integer k. Then the exponential factor becomes $\exp(i\omega_m t_k)$. We can also reverse the order of terms in the sum so that m ranges from its smallest value $(N+1)/2$ to its largest value $N-1$. The result is

$$\sum_{n=1}^{(N-1)/2} A_{N-n} e^{-i\omega_n t_k} = \sum_{n=(N+1)/2}^{N-1} A_n e^{i\omega_n t_k}, \tag{23.26}$$

where we have also replaced the "dummy" summation index m with the letter n.

Use this result (23.26) to replace the sum in the last term of Eq. (23.24). The three terms in S_k can be combined into a single sum from $n = 0$ to $n = N - 1$:

$$S_k = \frac{1}{N} \sum_{n=0}^{N-1} A_n e^{i\omega_n t_k}. \tag{23.27}$$

This is the inverse DFT in complex form. Notice that this formula has the same form as the DFT itself, Eq. (23.17), apart from the sign of the exponent and the overall factor of $1/N$.

The squared magnitude of the complex Fourier amplitude is $|A_n|^2 = A_n A_n^* = (C_n^2 + D_n^2)N^2/4$. This allows us to write the power spectrum (23.13) as

$$P_0 = \frac{1}{N^2}|A_0|^2, \tag{23.28a}$$

$$P_n = \frac{2}{N^2}|A_n|^2, \quad n = 1, \ldots, (N-1)/2, \tag{23.28b}$$

for odd N.

Exercise 23.3b

Use your results for A_n from the previous exercise to compute the inverse transform (23.27). Do your results match the original signal? Compute the power spectrum (23.28). Show that the average power (23.11) is equal to the total power $\sum_n P_n$.

One advantage of the complex form of the discrete Fourier transform is that we no longer need to restrict N to be odd. The definitions (23.17) and (23.27) for the DFT and its inverse are valid for any positive integer N, even or odd.

When N is even we need to modify the power spectrum. For even N the identity (23.23) tells us that the modes with $n = 1, \ldots, N/2-1$ are the complex conjugates of the modes with $n = N/2+1, \ldots, N-1$. These pairs of modes appear together, with a factor of 2, in the power spectrum. The zero mode $\omega_0 = 0$ and the highest frequency mode $\omega_{N/2} = \pi/\Delta t$ contribute separately. Thus, the power spectrum becomes

$$P_0 = \frac{1}{N^2}|A_0|^2, \tag{23.29a}$$

$$P_n = \frac{2}{N^2}|A_n|^2, \quad n = 1, \ldots, N/2-1, \tag{23.29b}$$

$$P_{N/2} = \frac{1}{N^2}|A_{N/2}|^2 \tag{23.29c}$$

when N is even.

Exercise 23.3c

Sample the signal $S(t) = (\sin(3t))^5$ at times $t_k = k\Delta t$ with $k = 0, \ldots, 99$ and $\Delta t = 0.1$. (Thus, $N = 100$.) Compute the discrete Fourier transform (23.17). Plot the real and imaginary parts of A_n as functions of angular frequency ω_n. (You might want to plot only the amplitudes with "small" indices, $i = 0$ through $i = N/2$.)

Exercise 23.3d

Using the amplitudes A_n from the previous exercise, compute the inverse discrete Fourier transform (23.27). Compare the real and imaginary parts of the result to the original sampled signal.

Exercise 23.3e

Plot the power spectrum for the signal $S(t) = (\sin(3t))^5$. Which frequencies have the most power? Can you explain why the power is concentrated at those frequencies?

23.4 Fast Fourier transform

The *fast Fourier transform* (FFT) is a very efficient algorithm for computing the discrete Fourier transform. A direct calculation of the amplitudes A_n from Eq. (23.17) requires a computation time proportional to N^2. The FFT reorganizes the calculation in a clever way so that the computation time is proportional to $N \log(N)$. For large N, this can be a huge savings.

The library `numpy.fft` includes a built-in FFT function called `fft()`. Given an array `S` containing the time domain amplitudes S_k, the command `A = numpy.fft.fft(S)` will create a complex array `A` containing the frequency domain amplitudes A_n. The command for the inverse FFT is `S = numpy.fft.ifft(A)`.

Exercise 23.4

Sample the signal $S(t) = (\cos(3t) - \sin(7t))^2$ at $N = 500$ times to generate discrete data t_k, S_k in the domain $0 \leq t \leq t_f$ with $t_f = 50$. (Note, $\Delta t = t_f/(N-1)$.) Use NumPy's FFT routine to obtain the discrete Fourier coefficients A_n. Compute the power spectrum and plot P_n versus frequency ω_n. Which frequencies contain the most power? Are these the expected values?

23.5 Nyquist frequency and aliasing

To keep the presentation as simple as possible, let's assume N is even. For a discrete signal S_k, the highest angular frequency in the power spectrum corresponds to $n = N/2$. This is called the *Nyquist critical frequency*,

$$\omega_{N/2} = \frac{\pi}{\Delta t}, \tag{23.30}$$

sometimes denoted ω_c.

Consider a continuous signal $S(t)$ with angular frequency exactly equal to $\omega_{N/2}$ for some sampling interval Δt. This signal has a period of $T = 2\pi/\omega_{N/2} = 2\Delta t$. When we sample this signal to obtain S_k there will be just two sample points per oscillation. This is the minimum number of sample points that can represent a wave—one sample point for each peak and one for each trough. If the signal $S(t)$ varies with angular frequency *greater* than $\omega_{N/2}$, the signal cannot be represented by the discrete values S_k.

For example, consider a sampling interval of $\Delta t = 1$. The Nyquist critical frequency is $\omega_{N/2} = \pi \approx 3.14$. Figure 23.1 shows a signal (solid line) with period $T = 4/3$ and angular frequency $\omega = 2\pi/T = 3\pi/2 \approx 4.71$. The sampling values S_k are represented by

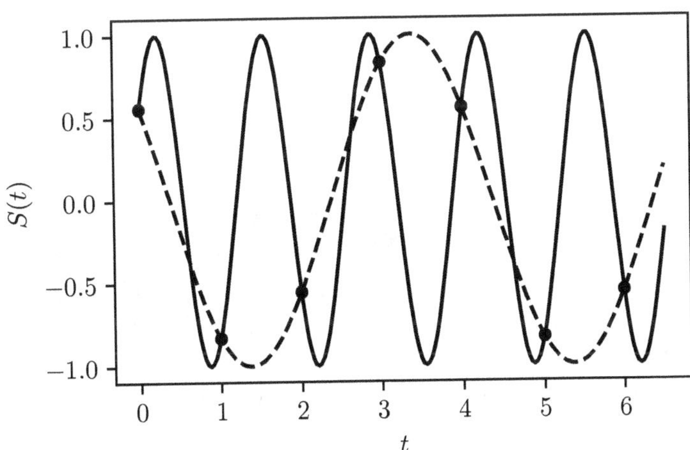

Fig. 23.1. The signal (solid line) has frequency above the Nyquist frequency. Discrete sampling (dots) causes this signal to be aliased to a lower-frequency signal (dashed line).

the heavy dots at intervals of $\Delta t = 1$. Because the signal frequency ($\approx$4.71) is greater than the Nyquist critical frequency ($\approx$3.14), the sample values S_k do not capture each peak and trough of the signal. The true signal (solid line) is misrepresented as a lower frequency signal (dashed line). We say that the signal has been *aliased* to a lower frequency. In this example, the signal is aliased to a frequency of $\pi/2 \approx 1.57$.

Exercise 23.5

Sample the signal $S(t) = \cos(5t)+\cos(10t)+\cos(15t)+\cos(20t)$ at N equally spaced times in the domain $0 \le t \le t_f$, where $t_f = 50$. This signal contains modes with frequencies $\omega = 5$, 10, 15, and 20. How does the Nyquist frequency depend on N? (Note that $\Delta t = t_f/(N-1)$.) Plot the power spectrum for various values of N with $\omega_{N/2} > 20$, and for values of N with $\omega_{N/2} < 20$. Where are the peaks in the power spectrum? Are they correct?

When computing the discrete Fourier transform, we must be cautious. If the Nyquist frequency is smaller than any of the frequencies contained in the signal, those frequencies will be aliased to lower values. The mode amplitudes A_n and power spectrum will misrepresent the true signal.

23.6 More Exercises

Exercise 23.6a

A typical high E guitar string has a mass per unit length of $\mu = 0.00039\,\mathrm{kg/m}$ and a tension of $T = 74.0\,\mathrm{N}$. The string extends from $x_a = 0$ to $x_b = L$, where $L = 0.66\,\mathrm{m}$. Solve the wave equation with initial conditions corresponding to a "pluck" with

$$y(0, x) = \frac{1}{100} e^{-100(x-L/2)^2} ,$$

and $v(0, x) = 0$. Use the NumPy FFT function to compute the discrete Fourier transform of the amplitude of the mid-point of the string, and find the power spectrum. What are the peak frequencies? The defining frequency for high E is $f = 329.63\,\text{Hz}$. (Recall that frequency f and angular frequency ω are related by $\omega = 2\pi f$.)

Exercise 23.6b

Simulate the driven, damped pendulum with "triple period" as in Exercise (19.4b). (The parameter values are $g = 9.8$, $m = 0.5$, $\ell = 0.11$, $\beta = 0.25$, and $\tau = 1.0$. Let $\gamma = 5.15$ and use initial conditions $\Theta(0) = \Omega(0) = 0.0$.) Carry out the simulation for $0 \le t \le 40$ with relatively high resolution. Compute the DFT with NumPy's FFT function, using the discrete numerical data for the angle $\Theta(t)$. Plot the power spectrum. What are the dominant frequencies? How are they related to the drive period τ? Note: The long-term behavior shows that Θ oscillates about $-\pi$. You can (partially) eliminate the zero-frequency mode by computing the DFT for $\Theta(t) + \pi$.

Appendix A

Data Files

The Python code below generates the data files used in the text. The data files are also available at the World Scientific Connect website (https://www.worldscientific.com/worldscibooks/10.1142/14375).

```
import numpy as np

##########################
# HubbleData.txt (chapter 12)
d = np.array([0.032,0.034,0.214,0.263,0.275,0.275,0.45,
              0.5,0.5,0.63,0.8,0.9,0.9,0.9,0.9,1.0,
              1.1,1.1,1.4,1.7,2.0,2.0,2.0,2.0])
v = np.array([170,290,-130,-70,-185,-220,200,290,270,
              200,300,-30,650,150,500,920,450,500,500,
              960,500,850,800,1090])
np.savetxt("HubbleData.txt",list(zip(d,v)))

##########################
# LineData.txt (chapter 12)
rnum = np.random.default_rng(4)
def f(x):
    return 3.6*x + 5.1

x = np.linspace(0,10,25)
y = f(x) + rnum.uniform(0,8,25)
np.savetxt("LineData.txt", list(zip(x,y)))
```

```
########################
# SineData.txt (chapter 12)
rnum = np.random.default_rng(5)
def f(x):
    return 2.3*np.sin(1.4*x + 0.5)

x = np.linspace(0,10,25)
y = f(x) + rnum.uniform(-1.2,1.2,25)
np.savetxt("SineData.txt", list(zip(x,y)))

########################
# PolyData.txt (chapter 12)
rnum = np.random.default_rng(3)
def f(x):
    return 0.8 - 0.31*x + .066*x**2 - .0038*x**3

x = np.linspace(0,10,25)
y = f(x) + rnum.uniform(-0.08,0.08,25)
np.savetxt("PolyData.txt", list(zip(x,y)))

########################
# InterpData.txt (chapter 12)
x = np.array([0.4,1.3,3.1,4.1,4.9,5.8,6.9])
y = np.array([1.7,2.9,2.8,2.0,1.1,0.7,1.3])
np.savetxt("InterpData.txt",list(zip(x,y)))

########################
# NonUniformData.txt (chapter 14)
rnum = np.random.default_rng(9)
def f(x):
    return np.sin(x) + np.cos(1.5*x)

x = rnum.uniform(-10,10,100) # random x values
x = np.sort(x)               # increasing order
y = f(x)
np.savetxt("NonUniformData.txt",list(zip(x,y)))

########################
# ToDifferentiateData.txt (chapter 16)
def f(x):
    return np.sin(3*x) + np.cos(5*x)

x = np.linspace(-np.pi,np.pi,101)
y = f(x)
np.savetxt('ToDifferentiateData.txt',list(zip(x,y)))
```

```
#########################
# SignalData.txt (chapter 23)
t = np.linspace(0,1,11)
S = np.array([-0.4,5.3,8.7,6.4,0.3,-2.6,-3.3,-5.4,5.2,2.4,1.6])
np.savetxt("SignalData.txt",list(zip(t,S)))
```

Index

S

www.ingramcontent.com/pod-product-compliance
Lightning Source LLC
Chambersburg PA
CBHW061624220326
41598CB00026BA/3870

* 9 7 8 9 8 1 9 8 1 6 6 4 4 *